STATISTICAL RESEARCH ON FATIGUE AND FRACTURE

Edited by

TSUNESHICHI TANAKA

Ritsumeikan University, Japan

SATOSHI NISHIJIMA

National Research Institute for Metals, Japan

MASAHIRO ICHIKAWA

University of Electro-Communications, Japan

Current Japanese Materials Research — Vol. 2

ELSEVIER APPLIED SCIENCE
LONDON and NEW YORK

ELSEVIER APPLIED SCIENCE PUBLISHERS LTD
Crown House, Linton Road, Barking, Essex IG11 8JU, England

Sole Distributor in the USA and Canada
ELSEVIER SCIENCE PUBLISHING CO., INC.
52 Vanderbilt Avenue, New York, NY 10017, USA

WITH 42 TABLES AND 142 ILLUSTRATIONS

© ELSEVIER APPLIED SCIENCE PUBLISHERS LTD and
THE SOCIETY OF MATERIALS SCIENCE, JAPAN 1987

British Library Cataloguing in Publication Data

Statistical research on fatigue and fracture.
— (Current Japanese materials research).
1. Materials — Fatigue — Statistical
methods   2. Fracture mechanics —
Statistical methods
I. Tanaka, Tsuneshichi   II. Nishijima,
Satoshi   III. Ichikawa, Masahiro   IV. Series
620.1'123'028   TA418.38

Library of Congress Cataloging in Publication Data

Statistical research on fatigue and fracture.

(Current Japanese materials research; vol. 2)
Bibliography: p.
Includes index.
1. Materials — Fatigue — Statistical methods.
2. Fracture mechanics — Statistical methods.   I. Tanaka,
Tsuneshichi, 1932-   .   II. Nishijima, Satoshi.
III. Ichikawa, Masahiro, 1940-   IV. Series.
TA418.38.S73   1987   620.1'123   87–5305

ISBN 1-85166-092-5

The selection and presentation of material and the opinions expressed are the sole responsibility of the author(s) concerned.

Phototypesetting by Tech-Set, Gateshead, Tyne & Wear.
Printed in Great Britain by the University Press, Cambridge.

Foreword

The Current Japanese Materials Research (CJMR) series is a new publication edited by the Society of Materials Science, Japan, and published by Elsevier Applied Science, UK, aiming at the overseas circulation of current Japanese achievement in the field of materials science and technology. This second volume of the series deals with *Statistical Research on Fatigue and Fracture* and follows the first volume on *Current Research on Fatigue Cracks*. All the papers have been selected to present the most important and substantial results obtained by the authors, in order to help readers to understand the different statistical approaches which have been used.

Although many international meetings are held every year in various specialized fields, it cannot be denied that most research results in Japan are published only in Japanese and tend therefore to be confined to the domestic audience. The publication of the CJMR series is an attempt to offer these results to colleagues abroad and thereby encourage the exchange of knowledge between us. I hope that our efforts will interest engineers and scientists in different countries and may contribute to the progress of materials science and technology throughout the world.

MORIYA OYANE
President, Society of Materials Science, Japan

Preface

Statistical approaches have been used for many years in various fields of research concerning materials science and technology. In particular, problems related to fatigue and fracture are often subject to this type of investigation, because both fatigue and fracture are known to be processes characterized by their stochastic nature.

Statistical research on fatigue and fracture has been an active area for the Fatigue Committee of the Society of Materials Science, Japan (JSMS). One of the recent achievements in this field is the compilation of two data bases on fatigue properties of metallic materials: one related to fatigue strength, prepared by collaboration between the Fatigue Committee and the Reliability Engineering Committee of JSMS, and published in 1982; the other on fatigue crack propagation, prepared by the Fatigue Committee, and issued in 1983. They include more than 2300 series of S-N type fatigue data and 900 sets of crack growth rate data for different metallic materials, provided by tests conducted since 1961 in various laboratories in Japan.

The co-operative work of the compilation and analysis of these data has encouraged the development of some new statistical techniques which may conveniently be used for similar computerized database systems. They include analytical techniques for small sample S-N data, criteria for the pooling of small sample data, analyses of pooled small samples, and so on.

It is also noted that the standard method of statistical fatigue testing was established at the Japan Society of Mechanical Engineers in 1981. More comprehensive fatigue data from a statistical point of view are expected to be accumulated in the future, as this standard procedure becomes popular in fatigue laboratories.

vii

In this volume, important progress in the field of probabilistic fatigue crack growth analysis is discussed, with the fracture mechanics approach being employed to varying degrees. The proposed models of life predictions and reliability analysis of structural components have required the use of the statistics-of-extremes technique.

Our basic understanding of life and/or strength distribution for different engineering materials, including various non-metals, has also been extended in recent years, and these results are now being constructively discussed between engineers and scientists with the aim of improving reliability of materials and structures.

Each of the chapters in this book represents a review of topical and important results achieved by the authors doing their statistical research on fatigue and fracture in Japan. The authors would welcome discussion of their work and the related problems with any reader interested in joining us to contribute to the advancement of materials science and technology.

TSUNESHICHI TANAKA
SATOSHI NISHIJIMA
MASAHIRO ICHIKAWA

Contents

Life Distribution and Assessment

List of Contributors

YASUMASA HAMAGUCHI
Second Airframe Division, National Aerospace Laboratory, 7-44-1 Jindaiji-Higashi, Chofu, Tokyo 182, Japan

MASAHIRO ICHIKAWA
Department of Mechanical Engineering, University of Electro-Communications, 1-5-1 Chofugaoka, Chofu, Tokyo 182, Japan

MINEAKI IIDA
Toshiba Electric Co. Ltd, 8 Shinsugita-machi, Isogo-ku, Yokohama, Kanagawa 235, Japan

HIROSHI ISHIKAWA
Department of Information Science, Kagawa University, 2-1 Saiwai-cho, Takamatsu 760, Japan

YOSHIYASU ITO
Hitachi Ibaraki Technical College, Hitachi Ltd, 2-17-2 Nishi Narusawa-cho, Hitachi 316, Japan

TAKASHI IWAYA
Numazu College of Technology, Ooka, Numazu, Shizuoka 410, Japan

HITOSHI KIMURA
Kagawa University, 1-1 Saiwai-cho, Takamatsu 760, Japan

HIDEO KITAGAWA
Department of Engineering, Yokohama National University, 153 Tokiwadai, Hodogaya-ku, Yokohama, Kanagawa 240, Japan

MAKOTO KITANO
Mechanical Engineering Research Laboratory, Hitachi Ltd, 502 Kandatsu, Tsuchiura 300, Japan

SHOTARO KODAMA
Department of Mechanical Engineering, Tokyo Metropolitan University, 2-1-1 Fukazawa, Setagaya-ku, Tokyo 158, Japan

TETSUO KUMAZAWA
Mechanical Engineering Research Laboratory, Hitachi Ltd, 502 Kandatsu, Tsuchiura 300, Japan

YUUJI NAKASONE
National Research Institute for Metals, Tsukuba Laboratories, 1-2-1 Sengen, Sakura-mura, Niihari-gun, Ibaraki 305, Japan

HIDETOSHI NAKAYASU
Materials System Research Laboratory, Kanazawa Institute of Technology, 7-1 Ogigaoka Nonoichi-machi, Ishikawa 921, Japan

HAJIME NAKAZAWA
Department of Mechanical Engineering, Chiba University, 1-33 Yayoicho, Chiba 280, Japan

SATOSHI NISHIJIMA
Fatigue Testing Division, National Research Institute for Metals, 2-3-12 Nakameguro, Meguroku, Tokyo 153, Japan

NAGATOSHI OKABE
Heavy Apparatus Engineering Laboratory, Toshiba Corporation, 1 Toshiba-cho, Fuchu 183, Tokyo, Japan. Present address: Heavy Apparatus Engineering Lab., Toshiba Corporation, 1-9, Suehiro-cho, Tsurumi-ku, Yokohama 230, Japan

TATSUO SAKAI
  Department of Mechanical Engineering, Faculty of Science and Engineering, Ritsumeikan University, Tojiin-Kitamachi, Kita-ku, Kyoto 603, Japan

TSUYOSHI SHIMAZAKI
  Graduate School, Yokohama National University, 153 Tokiwadai, Hodogaya-ku, Yokohama, Kanagawa 240, Japan

TASUKU SHIMIZU
  Mechanical Engineering Research Laboratory, Hitachi Ltd, 502 Kandatsu, Tsuchiura 300, Japan

TOSHIYUKI SHIMOKAWA
  Second Airframe Division, National Aerospace Laboratory, 7-44-1 Jindaiji-Higashi, Chofu, Tokyo 182, Japan

TSUNESHICHI TANAKA
  Department of Mechanical Engineering, Faculty of Science and Engineering, Ritsumeikan University, Tojiin-Kitamachi, Kita-ku, Kyoto 603, Japan

AKIRA TSURUI
  Department of Applied Mathematics and Physics, Kyoto University, Yoshida-honmachi, Sakyo-ku, Kyoto 606, Japan

NAMIO URABE
  Technical Research Center, Nippon Kokan KK, 1-1 Minamiwatarida-cho, Kawasaki-ku, Kawasaki 210, Japan

A. TOSHIMITSU YOKOBORI JR
  Department of Mechanical Engineering II, Tohoku University, Aoba, Aramaki, Sendai 980, Japan

TAKEO YOKOBORI
  Professor Emeritus of Mechanical Engineering, Tohoku University, Aoba, Aramaki, Sendai 980, Japan

Statistical Analysis of Small Sample Fatigue Data

SATOSHI NISHIJIMA
Fatigue Testing Division, National Research Institute for Metals,
2-3-12 Nakameguro, Meguroku, Tokyo 153, Japan

ABSTRACT

Although statistical fatigue data are recognized to be important for reliability based design of mechanical structures, it is not easy to obtain data for various materials with a large enough number of samples. This paper introduces a new statistical technique to analyze ordinary S-N data using a small number of samples while presuming no functional relationship between S and N other than the statistical distribution of the fatigue strength. The Probit analysis method with weighting is applied under an assumed coefficient of variation to estimate the mean fatigue strength at the observed fatigue life. Distribution of the relative strengths against the estimated mean strengths is then analyzed to deduce a new estimate for the coefficient of variation. The estimated mean strengths for different life levels may be fitted to a bi-linear or hyperbolic equation so as to resolve the S-N data into several parameters: slope, fatigue limit, knee position, roundness at the knee, and coefficient of variation in fatigue strength. Some additional remarks are given to rationalize the regression analysis of the S-N curve.

INTRODUCTION

There are increasing demands from engineers for data concerning statistical fatigue properties of various materials under different conditions to assess the reliability of machines and structural components. It is not easy, however, to obtain such data by carrying out fatigue tests with a large enough number of samples. Only very limited data are known to have been analyzed by conventional statistical analysis techniques. In contrast to this, there is a large amount of fatigue data sets each comprising small numbers of samples. The problem is how to extract statistical information from these small sample fatigue data.

The linear regression analysis method is prescribed in the ASTM standard E739-80 [1] for interpreting S-N type data, ignoring suspended

1

or run-out data as well as data of a non-linear nature, such as in the high-cycle regime. The standard method of statistical fatigue testing, S002-81 by JSME [2], stipulates the linear regression analysis for the inclined part of the $S-N$ curve and the staircase (up and down) method with several specimens for the horizontal part, presuming that the $S-N$ curve has a simple bi-linear shape. With the help of these two standards, one can deduce from the small sample fatigue data the confidence limits of the fatigue properties for given probabilities of failure, always under the hypothesis that the data can be fitted by a straight line.

The author proposed previously [3, 4] a method of statistical analysis to deduce the $P-S-N$ relation considering the distribution in the fatigue strength. The distribution of the relative strength deviation values was computed against the mean $S-N$ curve, which was determined by the Probit analysis method using weighting [5] at various fatigue life levels with small intervals. This distribution was regarded as that of the fatigue strength of the material and was then used to develop the $P-S-N$ relation. With this approach the $P-S-N$ relation can be obtained up to 1% of failure probability with 100 specimens in total, because the distribution is analyzed for pooled deviation values. Detailed explanation and applications for a wide range of materials with smooth and notched specimens have been described in other reports [4, 6–9].

This paper introduces a new method of analyzing ordinary $S-N$ data with a small number of samples [10, 11]. It is based on the Probit method, but the required number of data points is reduced to ordinary levels, around ten. This new method permits suspended data to be taken into account without assumptions about the shape of the $S-N$ curve, but presuming only a continuous distribution in the fatigue strength, which is a function of the fatigue life levels. Additional remarks will be given about the idea of fitting the analyzed results to a bi-linear or hyperbolic curve, so as to express the data with a few parameters: fatigue limit, slope, knee position, roundness at the knee, and coefficient of variation in fatigue strength. This method is believed to be used conveniently for compiling the fatigue data in database systems.

THEORY

Probit Analysis Method Using Weighting

The Probit analysis, as is conventionally employed in the field of fatigue research, is based on the assumption that the fatigue strength of a material is distributed according to the Gaussian distribution. Let S_m be the mean and v the coefficient of variation in the fatigue strength distribution. S_m and v can be determined from experimental response data comprising more than two sets of (S_i, p_i) values, where S_i is the test stress level and p_i is the observed probability of failure at S_i. The value of p_i is neither zero nor unity for the data to be valid. Necessary formulation will be as in the following:

$$P(u) = \int_{-\infty}^{u} Z(u)\, du \tag{1}$$

$$Z(u) = (1/\sqrt{2\pi})\exp(-u^2/2) \tag{2}$$

$$u = (1/vS_m)S - 1/v \tag{3}$$

where $P(u)$ is the cumulative probability, $Z(u)$ is the density and u is the normalized fatigue strength.

Equation (3) gives a linear relation providing a straight line as in Fig. 1 with a slope of $1/vS_m$ and an intersection with the ordinate of $-1/v$. The estimation of the two parameters, S_m and v, is possible by the ordinary least-squares method after converting the observed p_i value to the linearized value of u_i through the inverse function of Eqn. (1). However, as p_i is an observed value containing certain error, u_i also implies a sampling error which depends on the expected probability P at the stress S_i.

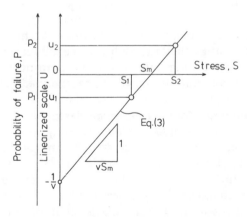

FIG. 1. Concept of Probit analysis.

In the least-squares calculation, therefore, the weighting should be considered as

$$w_i = n_i Z^2 / P(1 - P) \tag{4}$$

where n_i is the number of specimens tested at stress S_i for which p_i is evaluated. This weight is proportional to n_i and takes a maximum value at $P = 0.5$ but decreases rapidly when P leaves 0.5; for example, w becomes $1/2$, $1/4$ and $1/8$ of its maximum value at $P = 0.1, 0.03$ and 0.01, respectively.

The calculation can be obtained by an iterative process starting with rough estimates of S_m and v. The data sets (S_i, p_i) are converted into values (S_i, u_i, w_i) to get the two estimates:

$$\widehat{S}_m = \frac{\Sigma wS}{\Sigma w} - \frac{\Sigma wS^2 - (\Sigma wS)^2 / \Sigma w}{\Sigma wuS - \Sigma wu \Sigma wS / \Sigma w} \cdot \frac{\Sigma wu}{\Sigma w} \tag{5}$$

$$\widehat{v} = (\Sigma wS / \widehat{S}_m - \Sigma w) / \Sigma wu \tag{6}$$

where subscripts are eliminated for simplicity. The estimates obtained, $\widehat{S}_m$ and $\widehat{v}$, are then used to recalculate more accurate estimates in a repeated manner. It is clear that at least two valid response data sets (S_i, p_i) are required for the estimation of these two parameters, with p_i being neither zero nor unity.

A rapid convergence is obtained if the value of u is corrected, before entering into Eqns. (5) and (6), according to the difference of observed p and expected P, as [5]

$$u = u_0 + \{p - P(u_0)\} / Z(u_0) \tag{7}$$

where u_0 is the value directly converted from p, or that predicted from the result of preceding calculation. One of the advantages of using this correction is that the data with $p = 0$ or 1 can also be taken into consideration in the analysis, as will be exemplified later.

Extension of the Probit Method for Small Sample Data

For ordinary small sample S-N data the requirement of two valid response data tends not to be realized at any arbitrary life level. Therefore the application of the Probit analysis has been thought impossible for small sample S-N data. This is because there are two unknown parameters in the regression model of Eqn. (3).

Let the coefficient of variation, v, be given in the case of small sample data. The regression analysis then draws the best fit straight line in

Fig. 1 having a fixed intersection of $-1/v$ with the ordinate. In this case the minimum required valid response data is only one, and Eqn. (5) becomes [12]

$$\widehat{S}_m = \Sigma wS^2/(v\Sigma wuS + \Sigma wS) \qquad (8)$$

and the standard error for the estimate $\widehat{S}_m$ is evaluated by

$$e = v\widehat{S}_m^2/\sqrt{\Sigma wS^2} \qquad (9)$$

It may be noted that there is another way of reducing the unknown parameters in the regression model, that of setting the standard deviation, s, as the known variable instead of using v. The best fit straight line is then determined with a fixed slope of $1/s$, and the estimate and standard error become, respectively,

$$\widehat{S}_m = (\Sigma wS - s\Sigma wu)/\Sigma w \qquad (10)$$

$$e = s/\sqrt{\Sigma w} \qquad (11)$$

In most cases, it is difficult to say whether the coefficient of variation or the standard deviation should be regarded as known. However, it is frequent in practice that the coefficient of variation is assumed to be known by analogy to similar materials, and Eqn. (8) is adopted therefore in the present paper. The two methods actually result in a difference small enough to be neglected from the engineering viewpoint [11].

Once the mean fatigue strength is estimated for any desired life level, the value chosen for v may be checked by analyzing the distribution of relative strength values for individual test data, as in the following.

Multiply Censored Order Statistics

Consider the relative strength of each specimen

$$r = S/S_m \qquad (12)$$

where S is test stress and S_m the mean strength of the population at the observed fatigue life. Assuming that the fatigue strength follows the Gaussian distribution with a constant coefficient of variation, v, for different life levels, the value of r is also assumed to follow the Gaussian distribution with a standard deviation of v and a mean of $r_m = 1$. In reality, however, some of the r values would be the suspended data and the set of r values should be analyzed as multiply censored statistics. The parameters, r_m and v, could be estimated by several methods, namely Johnson's order statistics [12], or Nelson's hazard analysis [13], or more sophisticated methods such as the maximum likelihood method. Johnson's method is used in this paper as it is simple and suited for small sample analysis.

Let M be the total number of r values which are rearranged in ascending order and each labeled with the order number, K, defined regardless of the failure or non-failure of the specimen. The cumulative probability, p, for the Kth specimen, which should be a failed one, representing the probability for specimens weaker than the Kth, can be evaluated by an approximation formula for the median rank:

$$p = (J - 0.3)/(M + 0.4) \tag{13}$$

where J is the rank value given by the following equation [12]:

$$J = J_0 + (M + 1 - J_0)/(M + 2 - K) \tag{14}$$

In this equation, J_0 is the rank value attributed to the failed specimen just before the Kth specimen under consideration, and $J_0 = 0$ for $K = 1$.

All the r values for failed specimens having been trimmed to a set of data (r_i, p_i), the estimates for r_m and v can then be obtained in a manner similar to that described above:

$$\hat{v} = \{\Sigma r^2 - (\Sigma r)^2/n\}/(\Sigma ur - \Sigma u \Sigma r/n) \tag{15}$$

$$\hat{r}_m = (\Sigma r - \hat{v}\Sigma u)/n \tag{16}$$

where u is the value converted from p, and n is the number of data.

When the value $\hat{v}$ thus obtained is far from the value initially assumed, the process should be repeated with a corrected v. When $\hat{r}_m$ is considerably different from unity, the hypothesis of the Gaussian distribution should be questioned.

APPLYING THE PROPOSED METHOD FOR SMALL SAMPLE S-N DATA

In this section the calculation procedure will be shown first to determine the mean fatigue strength, S_m, at the observed fatigue life, N_j, under the given coefficient of variation, v, in the fatigue strength, and then to evaluate the distribution of the relative strength, r, to check the appropriateness of the selected value of v. For this purpose an example of the S-N data is used, as shown in Fig. 2. In total ten specimens are tested at four stress levels, resulting in eight specimens failing and two being suspended.

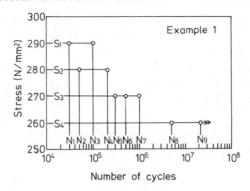

FIG. 2. Typical small sample S-N data; Example 1.

Estimating the Mean Fatigue Strength at the Observed Life

It is necessary to consider the variations of failure probability, p, at each stress level according to the change in the N values. A typical result is given in Table 1 using the data in Example 1 in Fig. 2. For example, just before N_5 cycles (denoted as N_5^-), $p = 1$ for S_1 and S_2 and $p = 0$ for S_3 and S_4; just after N_5 cycles (denoted as N_5^+) the value for S_3 is changed to $p = 1/3$, and just at N_5 cycles p may be evaluated by Eqn. (13) as $p = 0.7/3.4 = 0.21$. As seen in Table 1, it is clear that p for given S_i never decreases as N_j increases and the row of p data at N_j^+ is the same as that at N_{j+1}^-.

TABLE 1
Distribution of failure probability for the data of Example 1

Stress (N/mm^2)	Number of cycles				
	N_5^-	N_5	N_5^+	N_6^+	N_7^+
$S_1 = 290$	1·00	1·00	1·00	1·00	1·00
$S_2 = 280$	1·00	1·00	1·00	1·00	1·00
$S_3 = 270$	0·00	0·21	0·33	0·67	1·00
$S_4 = 260$	0·00	0·00	0·00	0·00	0·00

It seems possible to estimate the mean strength at N_5, $S_m(N_5)$, using p data at N_5 in Table 1 while referring to Eqn. (8), under a given value of v. However, this can be misleading in the re-evaluation of v. In fact, the smaller the values chosen for v, the closer the calculated S_m would be to S_3, because the regressed line should pass the plot for S_3 on Fig. 1 as there is no other valid data. Similarly, if the valid data are always only one for other N_j levels, all the r values would be close to unity, and the new estimate of v would decrease even more.

To avoid this problem, $\widehat{S}_m(N_j)$ is defined as the weighted mean of $\widehat{S}_m(N_j^-)$ and $\widehat{S}_m(N_j^+)$, considering the change of p values before and after N_j. The estimated mean strength and standard error of the estimate are respectively

$$\widehat{S}_m(N_j) = \frac{\widehat{S}_m(N_j^-)e^2(N_j^+) + \widehat{S}_m(N_j^+)e^2(N_j^-)}{e^2(N_j^-) + e^2(N_j^+)} \tag{17}$$

$$e(N_j) = \frac{e(N_j^-)e(N_j^+)}{e^2(N_j^-) + e^2(N_j^+)} \tag{18}$$

where $\widehat{S}_m$ and e at N_j^- are of course equal to those at N_{j-1}^+. As for $\widehat{S}_m(N_1)$, the p values at N_1 are used directly because no data is available for N_1^-; similarly $\widehat{S}_m$ for maximum cycles, N_{max}, is defined as $\widehat{S}_m(N_{max}^-)$.

Table 2 gives the results of calculations for the data in Example 1 under a value for v of 0·02. It is to be noted that the mean strength can be evaluated even for those cases without valid data, i.e. with p neither zero nor unity, such as for N_4^+, N_7^+ and N_8^+ in Table 2. This is the effect of utilizing Eqn. (7), as mentioned above.

TABLE 2
Analytical results for the data of Example 1 ($v = 0·02$)

N_i	$S_m(N_i^+)$	$e(N_i^+)$	$\widehat{S}_m(N_i)$	$e(N_i)$
$N_1 = 3 \times 10^4$	290·3	4·9	293·3	5·4
$N_2 = 5 \times 10^4$	285·0	3·6	286·9	2·9
$N_3 = 1 \times 10^5$	280·1	4·3	283·0	2·8
$N_4 = 2 \times 10^5$	275·7	5·0	278·2	3·3
$N_5 = 3 \times 10^5$	272·0	3·7	273·3	3·0
$N_6 = 5 \times 10^5$	268·2	3·6	270·0	2·6
$N_7 = 1 \times 10^6$	264·9	4·5	266·9	2·8
$N_8 = 5 \times 10^6$	261·7	3·6	263·0	2·8
$N_9 = 2 \times 10^7$	—	—	261·7	3·6

Figure 3 illustrates the results: the mean strength response is expressed in a stepwise curve and the estimated mean strength at each observed life as points along this curve. It is now possible to check the value chosen for v by analyzing the distribution of r values.

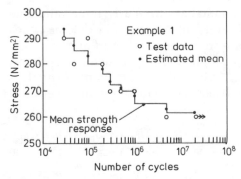

FIG. 3. Mean strength response for the data of Example 1.

Evaluation of the Coefficient of Variation in Fatigue Strength

Table 3 gives calculated values of the relative strength, r, for each specimen arranged in order of magnitude and cumulative probability, p, after Eqns. (13) and (14). The estimates obtained from Eqns. (15) and (16) are $\hat{r}_m = 1 \cdot 002$ and $\hat{v} = 0 \cdot 0198$ against values initially assumed to be $r_m = 1$ and $v = 0 \cdot 02$; thus these results may be accepted as final.

TABLE 3
Relative fatigue strength for the data of Example 1

No.	Stress (N/mm^2)	Cycle	Relative strength	Rank value	Cumulative probability
K	S	N	r	j	p
1	280	5×10^4	0·982	1·000	0·067
2	270	3×10^5	0·988	2·000	0·163
3	260	5×10^6	0·989	3·000	0·260
4	290	3×10^4	0·989	4·000	0·356
5	260	$>2 \times 10^7$	>0·994	—	—
6	260	$>2 \times 10^7$	>0·994	—	—
7	270	5×10^5	1·000	5·400	0·490
8	280	2×10^5	1·006	6·800	0·625
9	270	1×10^6	1·012	8·200	0·760
10	290	1×10^5	1·035	9·600	0·894

Other Examples of Application

Figure 4 shows typical small sample data, obtained for carburized Cr–Mo steel under rotating bending tests, where six specimens in total are tested at five stress levels with two runouts. As seen in this figure,

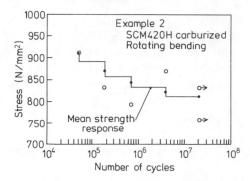

FIG. 4. Data for case-hardened steel; Example 2.

while the proposed small sample Probit method can successfully plot the mean strength response, it would be almost impossible by the conventional eye-fitting method even to read the average tendency of the data. The coefficient of variation in fatigue strength is evaluated as 0·07 in this case, giving a reference for comparison with other data sets.

Figure 5 is another example extracted from a series of test results at elevated temperatures. There seem to be two groups of data: those in the low-cycle range and those in the high-cycle range. A similar effect is often observed in corrosion fatigue or in fatigue of hard materials. It is clear that the present analysis is the most appropriate to study this unusual behavior because it assumes no fixed shape for the S–N curve.

It may be noted that the present analysis is based on the hypothesis of a constant coefficient of variation in fatigue strength for different life

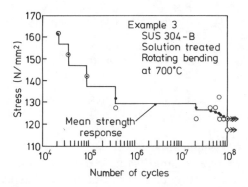

FIG. 5. Data for high-temperature fatigue; Example 3.

levels. The analysis can be consistent only when the object data belong to a single failure mechanism. This point is disregarded in Fig. 5, as there is not sufficient replication of data to be analyzed in the low-cycle range. The deduced coefficient of variation is 0·027, which may be considered for this case to be due almost entirely to the high-cycle failure mechanism.

PARAMETRIC REPRESENTATION OF *S-N* DATA

As stated above, the proposed method of statistical analysis does not imply a fixed shape for the *S-N* curve. It is thought however that the mathematical expression for the *S-N* curve could be useful especially in comparing and storing numerous data sets for engineering purposes. In fact, a linear equation on semi-log or log-log scales is frequently used [1, 2, 14], but it cannot express the whole *S-N* curve including the fatigue limit.

Typical mathematical forms for the *S-N* relation, including the fatigue limit, are compared in Fig. 6, in which the lower right is the model proposed in this work. As the four models are all non-linear and irrational functions, it is not simple to operate them for the optimization of the unknown parameters in the formula. Nevertheless, the proposed

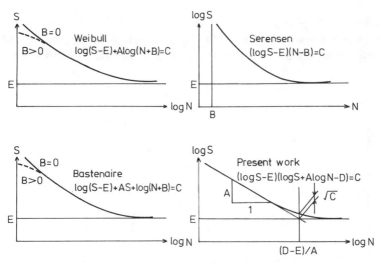

FIG. 6. Typical formulations for the *S-N* curve.

model reveals some advantages over the others in the possibility of resolving the $S-N$ data into four parameters: slope, knee position, fatigue limit, and sharpness at the knee. The coefficient of variation in fatigue strength will be the fifth parameter to characterize a set of $S-N$ data.

The proposed $S-N$ curve may be expressed as

$$y = a\{\sqrt{(x - .d)^2 + c^2} - x\} + b \qquad (19)$$

where $y = \log S$, $x = \log N$, and a, b, c and d are constants to be optimized. When $c = 0$ this equation yields the simplest bi-linear relation composed of inclined and horizontal straight lines; the slope is then given by $-2a$, the knee position by d, and the fatigue limit by $b - ad$.

The fitting of Eqn. (19) to the experimental data can be done by the ordinary least-squares method since the $S-N$ data including runouts are already converted to the $\widehat{S}_m - N_j$ data set with standard error e for each $\widehat{S}_m$ value. Thus the weighted sum of squared residuals, V, is to be minimized by an iterative process:

$$V = \Sigma\Sigma w(\log \widehat{S}_m - y)^2 \qquad (20)$$

where

$$w = (\ln 10)^2/\ln(e^2/\widehat{S}_m^2 + 1) \qquad (21)$$

Example of Application

Figure 7 shows a bi-linear fitting of the data in Example 1. Mean fatigue strengths at each observed life are plotted as points, and the

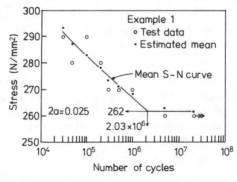

FIG. 7. Fitting of the proposed $S-N$ curve to the data for Example 1.

fatigue limit is defined as the lowest of these points. It is possible to evaluate the coefficient of variation in fatigue strength against the fitted mean S-N curve, which would be used to develop the P-S-N relation. The value calculated for this case is 0·0187, slightly different from that evaluated without fitting the mean S-N curve, where the difference can be ignored from an engineering point of view.

Figure 8 illustrates another example, which has no obvious knee point but nicely fits the proposed model. The number of cycles for the knee point was 2·36 × 10⁴, corresponding to $d = 4·372$, so all the data were regarded as being beyond the assumed knee point.

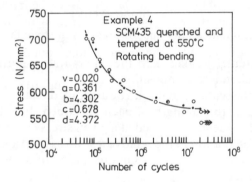

FIG. 8. Fitting of the proposed S-N curve to the data for Example 4.

Alternative Method of Fitting the Bi-linear S-N Curve

If the problem is only to fit a bi-linear curve to a set of S-N data, the analysis of $\hat{S}_m(N_j)$ described above is not really necessary, and an alternative method based on regression analysis may be adopted. In fact the data belonging to the inclined part of the S-N curve may easily be fitted to a straight line. The mean square error to the fitted line may be borrowed to evaluate the standard deviation or the coefficient of variation in fatigue strength, which is subsequently used to estimate the mean fatigue limit according to the small sample Probit method. In this process, the knee point is selected iteratively so as to minimize the mean square error determined against the obtained bi-linear S-N curve.

However, there is an important problem in treating the S-N data in this manner. Figure 9 shows a typical data scatter on a heat-treated medium carbon steel, extracted from the NRIM Fatigue Data Sheet [15]. The alphabetic codes in the figure designate the differences in the bar stock from which the specimens were prepared, showing bars B and

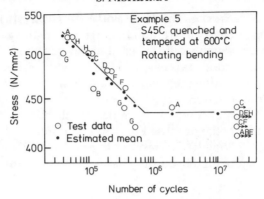

FIG. 9. Typical data scatter due to variations in strength of specimens.

G to be apparently weaker than the others due seemingly to a slight difference in chemical composition.

In general, the scatter of data can be attributed relatively frequently to the difference in the strength of each specimen. In other words, the scatter in the fatigue life observed at a given stress can be considered to be due to the fluctuation in the relative test stress level against the actual strength of each specimen. From this point of view, the ordinary regression model

$$X = b - aY \qquad (22)$$

where $X = \log N$, $Y = S$ or $\log S$, and a and b are constants, does not seem to be the most adequate, because it implies that the cause of scatter lies uniquely in X.

When the cause of scatter is attributable to both X and Y, the model for the regression should be written as

$$\alpha X + \beta Y = p \qquad (23)$$

where α, β and p are constants. In the case of Eqn. (22) the constants are determined so as to minimize the sum of squared residuals in the X direction; in the case of Eqn. (23) the residuals are to be evaluated in the direction perpendicular to the regressed line. Here the scale for X and Y should be standardized according to the span of data along each axis, so that the values of perpendicular residuals are unique:

$$X_0 = (X - \bar{X})/\sqrt{S_{xx}/(n-1)}$$
$$Y_0 = (Y - \bar{Y})/\sqrt{S_{yy}/(n-1)}$$

Where n is the number of samples, $\overline{X}$ and $\overline{Y}$ are the means for X and Y, respectively, and

$$S_{xx} = \Sigma X^2 - (\Sigma X)^2/n$$

$$S_{yy} = \Sigma Y^2 - (\Sigma Y)^2/n$$

In fact, this is the same treatment as in the principal component analysis; the slope of the optimized straight line is known to be ± 1 when the span of data is equal along both axes. The solution in this case is written as

$$X_0 - Y_0 = 0$$

and the constants in the form of Eqn. (22) are

$$\widehat{a} = \sqrt{S_{xx}/S_{yy}} \tag{24}$$

$$\widehat{b} = X + \widehat{a}Y \tag{25}$$

This method is called 'simultaneous regression' in the present paper.

Figure 10 demonstrates for sample data the results of the three different regression methods:

(1) $X = b - aY$

(2) $Y = d - cX$

(3) $\alpha X + \beta Y = p$

The result by simultaneous regression coincides with the median plots and also with that by the Probit method, whereas the other two

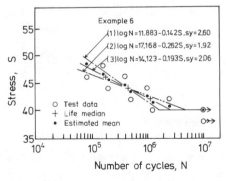

FIG. 10. Linear curve fitting with three different models.

methods cannot be said to represent the data. The solid line in Fig. 9 is also computed by the simultaneous regression method, presenting good agreement with the plots by the Probit method.

EXTENSION OF THE PROBIT METHOD TO THE OTHER DISTRIBUTIONS

Since its introduction, the Probit method has been used for analyzing data where the Gaussian distribution is presumed. However, the distribution shape itself is not of the first importance for this method, as can be understood from Fig. 1 or Eqns. (1)–(3). For fatigue data, it is not rare that the distribution in strength is skewed with a larger scatter in the weaker side, as for example in the case of notched specimens of steels [3, 6, 7] or smooth specimens of some aluminium alloys in the high-cycle regime [3, 8]. In such cases where the Gaussian distribution is judged to be apparently misleading, asymmetric distributions can be adopted with the Probit method. Table 4 gives some necessary formulations regarding the logistic and Weibull distributions as examples, together with the case of Gaussian distribution for comparison.

The assumption of a known coefficient of variation, which has an equivalent meaning to a known standard deviation on the logarithmic scale, accomplishes nothing but to reduce the unknown parameters in the model to be applied to the small sample data. It is therefore irrelevant whether we choose either a or b as known for distributions in Table 4. By analogy to the Gaussian distribution, however, it would be better to assume the shape parameter a for the logistic and Weibull distributions to be known.

It may be noted that the statistical test of fitness to a certain distribution shape is in general useless for a limited number of samples. The proper choice of distribution function is better achieved by referring to experience with a larger number of data or by engineering judgment.

CONCLUDING REMARKS

There are strong demands by engineers for statistical fatigue data on various materials for the purpose of reliability based designing. However, the published data with a large enough number of samples

TABLE 4
Extension of Probit analysis for different distributions

	Gaussian distribution	Logistic distribution	Weibull distribution
Distribution function	$P(u) = \int_{-\infty}^{u} Z(u)\,du$	$P(u) = \dfrac{1}{1 + e^{-u}}$	$P(u) = 1 - \exp(-e^u)$
Density function	$Z(u) = \dfrac{1}{\sqrt{2\pi}} \exp\left(-\dfrac{u^2}{2}\right)$	$Z(u) = \dfrac{e^{-u}}{(1 + e^{-u})^2}$ $= P(1-P)$	$Z(u) = e^u \exp(-e^u)$ $= (P-1)\ln(1-P)$
Weight	$w(u) = \dfrac{nZ^2}{P(1-P)}$	$w(u) = \dfrac{ne^u}{(e^u + 1)^2}$ $= nP(1-P)$	$w(u) = \dfrac{ne^{2u}}{\exp(e^u) - 1}$ $= n\dfrac{1-P}{P}\{\ln(1-P)\}^2$
Inverse function	$U(P) = P^{-1}(P)$	$U(P) = \ln\dfrac{1}{1-P}$	$U(P) = \ln\left(\ln\dfrac{1}{1-P}\right)$
Regression model	$u = ax + b$ $\hat{\sigma} = \dfrac{1}{\hat{a}}, \;\; \mu = -\dfrac{\hat{b}}{\hat{a}}$	$\hat{\mu} = ax + b$	$u = a\ln x + b$ $\hat{\gamma} = \hat{a}, \;\; \hat{\alpha} = e^{-\hat{b}}$

applicable for this purpose are very limited. In contrast to this, there exists a large amount of ordinary fatigue data with a relatively small number of samples. The aim of the present paper is to present some ideas for extracting statistical information from these ordinary small sample fatigue data.

A new method is proposed for the analysis of small sample S–N data presuming no functional relationship except for the statistical property of the data scatter. The Probit analysis method using weighting is modified for application to a limited number of samples; a coefficient of variation in fatigue strength is first assumed to determine only the mean strength at different life levels, then the assumed value is checked by analyzing the pooled distribution of relative strength deviation values for each datum against the estimated mean.

On the other hand, the estimated mean strength responses for different life levels are conveniently fitted to a bi-linear or a hyperbolic curve, so that a set of S–N data can be resolved into several parameters: slope, knee position, fatigue limit, sharpness at the knee, and coefficient of variation in fatigue strength. An alternative method for this is made possible by combining a linear regression technique and the modified Probit method.

It is pointed out that the regression model would be more consistent when considering the squared residuals in the direction perpendicular to the regressed line in the case of S–N data, which can be achieved by a method called 'simultaneous regression' in this paper.

REFERENCES

[1] *Standard Practice for Statistical Analysis of Linear or Linearized Stress-Life (S-N) and Strain-Life (ε-N) Fatigue Data*, ASTM Standard, E739-80, American Soc. Testing and Materials (1980).

[2] *Standard Method of Statistical Fatigue Testing*, JSME S002-81, Japan Soc. Mechanical Engineers (1981).

[3] S. Nishijima, Mechanical behavior of materials, *Proc. 1974 Symp. Mech. Behav. Materials*, Kyoto, Japan Soc. Materials Science, **1**, 417–26 (1974).

[4] S. Nishijima, *Statistical Analysis of Fatigue Data*, STP 744, American Soc. Testing and Materials, 75–88 (1981).

[5] E. S. Pearson and H. O. Hartley, *Biometrika Tables for Statisticians*, Cambridge Univ. Press, New York, **1**, 4 (1962).

[6] S. Nishijima *et al.*, *Trans. NRIM*, National Research Inst. for Metals, Tokyo, **19**, 119–32 (1977).

[7] S. Nishijima *et al.*, *Trans. NRIM*, **19**, 327–43 (1978).

[8]   S. Nishijima *et al.*, *Trans. NRIM*, **20**, 314–20 (1979).
[9]   S. Nishijima and E. Takeuchi, *Trans. NRIM*, **21**, 74–84 (1979).
[10] S. Nishijima, *J. Soc. Mater. Sci., Japan*, **29**, 24–9 (1980).
[11] S. Nishijima, *Trans. Japan Soc. Mech. Engrs.*, **47**, 1303–13 (1980).
[12] L. G. Johnson, *The Statistical Treatment of Fatigue Experiments*, Elsevier, New York, **37** (1964).
[13] W. Nelson, *J. Qual. Technol.*, **1**(1), 27–52 (1969).
[14] R. E. Little and E. H. Jebe, *Statistical Design of Fatigue Experiments*, Applied Science Publishers, London (1975).
[15] *Data Sheets on Fatigue Properties of S45C (0·45C) Steel for Machine Structural Use*, National Research Inst. for Metals, Tokyo, 14pp. (1978).

Method of Pooling Fatigue Data and Its Application to the Data Base on Fatigue Strength

HIDETOSHI NAKAYASU

Materials System Research Laboratory, Kanazawa Institute of Technology, 7-1 Ogigaoka Nonoichi-machi, Ishikawa 921, Japan

ABSTRACT

A method of pooling data in the fatigue data base is proposed. The procedure consists of a statistical data pooling of two kinds of fatigue data and the application of this to multiple samples from the data base. This data pooling method is based on a statistical test of the significant difference between the regression equations fitted to the $S-N$ data. There are six kinds of conditions required for the method of pooling data: (i) statistical evaluation for inclined parts of the $S-N$ data: test of linearity of data (condition 1); test of equality of residual squares between regression equation and data (condition 2); test of equality of regressive coefficients (conditions 3 and 4); (ii) statistical evaluation for horizontal parts of the $S-N$ data: test of equality of mean values (condition 5); test of equality of variances (condition 6). An algorithm for pooling the multiple samples is also developed as an extension of the method. The proposed method is applied to the data base on fatigue strength of metallic materials compiled by the Japanese Society of Materials Science. An application of the method to reliability-based fatigue design is described.

INTRODUCTION

It is recognized that as the number of kinds of functions required in structural design increases, the higher will be the required level of reliability and safety of the structure. As structural systems are subjected to various environmental conditions, a great number of experimental data are required for an understanding of the fatigue properties of each structural element. Nevertheless, it is difficult to store these fatigue data separately for specific conditions. Therefore many research institutes store the data together in a composite data base to utilize multi-objective fatigue design of structures [1-4]. Figure 1 shows the general features for describing the relationship between reliability-based design and this

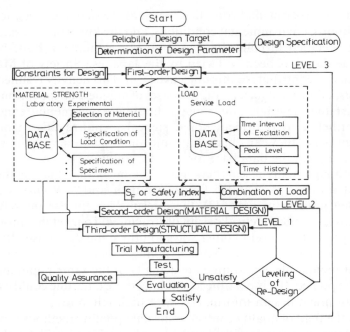

FIG. 1. General features of the process of reliability-based design concerned
with the data base.

data base. This figure is a flow-chart of the proposed design process
based on a data base system and on the concept that the target of
reliability-based design kept within the constraints of economy and
quality assurance should be a controlling factor in the decision-making
process for design and construction of all major structural systems. In
this flow-chart there are accordingly three feed-back loops corresponding to the levels of re-design, i.e. the first-order (design specification),
the second-order (material design), and the third-order (structural
design).

Though it seems that an effective data base concerned with fatigue
strength of materials has been compiled by the public institutes and
academic associations, there remain problems which must be solved in
order to utilize this data base effectively for the practical design of
structures. One of these problems is that mechanical designers do not
find it useful in their design of a structure. This is because it offers little
practical or useful information concerning their design purpose when

restricted experimental conditions, such as the shape of specimen, load conditions, environmental conditions, and the kinds of materials, have been specified in collecting the data. For example, a large body of fatigue data has been collected by the Japanese Society of Materials Science to contribute to this objective [3, 4]. This activity resulted in the construction of a data base of 2317 cases of fatigue data for various kinds of metallic materials under many different experimental conditions. Nevertheless, it is difficult for designers to utilize this in practical designs since they can obtain little information in line with their design purpose from the data base for specific conditions.

Although many researchers have developed analytical methods for small samples to evaluate fatigue strength and S-N curves from a limited number of data [12], these results do not use the data base effectively in the practical design of structures. To meet this problem, it is important to develop other ways to utilize the data base efficiently. For example, it is a reasonable hypothesis that data differing only in experimental place and time are actually of the same population and can be pooled. If it is possible to combine these data, one could expect to obtain much design information with high reliability.

In the field of quality control, some data pooling methods have been developed concerned with mean value and variance [5–7]. On the other hand, a standardized method for judging the equality of two S-N curves has been suggested by the Japanese Society of Mechanical Engineers (JSME) [8]. However, these methods are not suitable to the treatment of multi-variable data or multiple samples in a data base, such as fatigue data. This paper deals with a simple and practical method for pooling data by the use of a graphical procedure capable of applying the multiple samples, such as fatigue data, in a data base. An application to the reliability design of a structure will also be illustrated using the fatigue data base.

IDENTIFICATION OF THE MODEL FOR FATIGUE DATA

Definition of Fatigue Data

Figure 2 shows two kinds of fatigue data where they are plotted on a log-log scale. In this figure, $S_{i(\cdot)}$ designates the value of the ith stress level, $l_{(\cdot)}$ the number of stress levels, $N_{ij(\cdot)}$ the jth number of cycles to failure on stress level $S_{i(\cdot)}$ ($i = 1, \ldots, l_{(\cdot)}, j = 1, \ldots, m_{i(\cdot)}$), and $m_{i(\cdot)}$ the number of specimens that failed on the ith stress level, $S_{i(\cdot)}$. The symbol

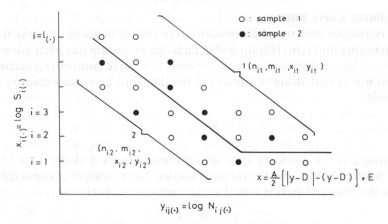

FIG. 2. Plots of fatigue data on a log-log scale.

$(\cdot)$ refers to each sample 1 and 2. From these definitions, the vertical and horizontal axes, $x_{i(\cdot)}$ and $y_{ij(\cdot)}$, are

$$x_{i(\cdot)} = \log S_{i(\cdot)}$$

and (1)

$$y_{ij(\cdot)} = \log N_{ij(\cdot)}$$

Therefore

$$n_{I(\cdot)} = \sum_{i=1}^{l_{(\cdot)}} m_{i(\cdot)}, \qquad \bar{x}_{(\cdot)} = \sum_{i=1}^{l_{(\cdot)}} m_{i(\cdot)} \cdot x_{i(\cdot)}/n_{I(\cdot)},$$

$$\bar{y}_{(\cdot)} = \sum_{i=1}^{l_{(\cdot)}} \sum_{j=1}^{m_{i(\cdot)}} y_{ij(\cdot)}/n_{I(\cdot)}$$

(2)

and

$$S_{xx(\cdot)} = \frac{1}{n_{I(\cdot)}} \sum_{i=1}^{l_{(\cdot)}} m_{i(\cdot)} \cdot (x_{i(\cdot)} - \bar{x}_{(\cdot)})^2,$$

(3)

$$S_{xy(\cdot)} = \frac{1}{n_{I(\cdot)}} \sum_{i=1}^{l_{(\cdot)}} (x_{i(\cdot)} - \bar{x}_{(\cdot)}) \sum_{j=1}^{m_{i(\cdot)}} (y_{ij(\cdot)} - \bar{y}_{(\cdot)})$$

where $n_{I(\cdot)}$ is the total number of specimens.

Bi-linear Curve Fitting

A simple and novel statistical curve fitting method developed by Nishijima and Ishii [13] for $S-N$ type fatigue test data has been adopted here. This method uses a bi-linear type $S-N$ curve comprising inclined and horizontal straight lines; thus the data can be represented by the model

$$x = \frac{A}{2}\{|y - D| - (y - D)\} + E \qquad (4)$$

which can be resolved into four parameters: A = slope, D = knee, E = fatigue limit, and the range of scatter. The principle of curve fitting for the inclined part is based on the regression analysis:

$$y = \alpha + \beta x + \varepsilon \qquad (5)$$

where

$$\widehat{\beta} = \frac{S_{xy}}{S_{xx}} \qquad (6)$$

$$\widehat{\alpha} = \overline{y} - \widehat{\beta}\overline{x} \qquad (7)$$

and for the horizontal part on the Probit analysis technique [13].

DATA POOLING ANALYSIS

Pooling Method for Two Kinds of Fatigue Data

The proposed method of pooling two kinds of fatigue data is based on the following criterion: are fitted curves to the fatigue data the same or not? Thus, the following conditions are criteria for judging the equality of the model fitted to each $S-N$ curve on the basis of a statistical significance test.

(i) Statistical Evaluation for Inclined Parts of S-N Data

(1) *Condition 1: test of linearity of samples.* The samples, which are plotted on the log-log scale, must satisfy the condition of linearity. The steps to test this linearity are in accordance with the standard method of statistical fatigue testing specified by JSME [8].

(2) *Condition 2: test of equality of residual squares between the regression equation and the data.* Samples must satisfy the statistical test of equality of residual squares between the original data and the regression model fitted to the data.

(3) *Condition 3: test of equality of the slope of the regression equations.*
Samples must satisfy the statistical test of equality of regression
coefficients that stand for the slope of two regression equations fitted to
the two sample data.

(4) *Condition 4: test of equality of the intercepts of the regression
equations.* Samples must satisfy the statistical test of equality of regression
coefficients that stand for the intercepts of the two regression
equations.

(ii) Statistical Evaluation for Horizontal Parts of S–N Data
The horizontal part, i.e. the fatigue limit, is determined by the Probit
analysis (up-and-down analysis) [13], where it is supposed that the
fatigue strength is distributed normally. Thus, in addition to the four
conditions for inclined parts, the following two conditions must be
checked for horizontal parts.

(5) *Condition 5: test of equality of means of fatigue limit.* The mean
values of fatigue limits for each sample must be identical. Thus, this
condition should satisfy the statistical test of equality of the mean values
of fatigue limits.

(6) *Condition 6: test of equality of variances of fatigue limit.* It is the 6th
condition to satisfy the statistical test of equality of variances of fatigue
limits.

(iii) Practical Calculation Procedure and Evaluation for Inclined Parts
(1) *Condition 1: test of linearity of samples.* This is the statistical test for
the hypothesis that the samples for inclined parts of S–N data plotted on
a log-log scale satisfy the condition of linearity. The steps to test the
hypothesis are in accordance with the standard method of statistical
fatigue testing by JSME [8]. If

$$F_0 < F(l - 2, n_I - l; \gamma) \tag{8}$$

is satisfied, the hypothesis can be accepted; otherwise the data cannot be
pooled. In Eqn. (8), when there are more than two measurements at each
stress level, F_0 is given by

$$F_0 = \frac{\sum_{i=1}^{l} m_i \cdot [y_{ij} - \{\bar{y} + \hat{\beta}(x_i - \bar{x})\}]^2/(l - 2)}{\sum_{i=1}^{l}\sum_{j=1}^{m_i} (y_{ij} - \bar{y}_i)^2/(n_I - l)} \tag{9}$$

where

$$n_I = \sum_{i=1}^{l} m_i, \qquad \bar{x} = \sum_{i=1}^{l} m_i \cdot x_i/n_I,$$

$$\bar{y} = \sum_{i=1}^{l} \sum_{j=1}^{m_i} y_{ij}/n_I, \qquad y_i = \sum_{i=1}^{l} \sum_{j=1}^{m_i} y_{ij}/m_i \qquad (10)$$

and $F(l - 2, n_I - 1; \gamma)$ in Eqn. (8) is a γ percentile point of F-distribution with the degrees of freedom $l - 2$ and $n_I - 1$. On the other hand, if there are not more than two measurements at each stress level, F_0 is calculated by

$$F_0 = \frac{S_R}{S_{yx}/(n_I - 2)} \qquad (11)$$

where

$$S_R = n_I \cdot S_{xy}^2/S_{xx}, \qquad S_{yx} = \sum_{i=1}^{l} \{y_i - (\hat{\alpha} - \hat{\beta}x_i)\}^2,$$

$$S_{xx} = \frac{1}{n_I} \sum_{i=1}^{l} (x_i - \bar{x})^2, \qquad S_{xy}^2 = \frac{1}{n_I} \left\{ \sum_{i=1}^{l} (x_i - \bar{x})(y_i - \bar{y}) \right\}^2$$

(2) *Condition 2: test of equality of residual squares between the regression equation and the data.*

(Step 1) Calculate the estimator of error variance $\hat{\sigma}_{(\cdot)}^2$ between measurements and the regression equation

$$\hat{\sigma}_{(\cdot)}^2 = \frac{1}{n_{I(\cdot)} - 2} \sum_{i=1}^{l(\cdot)} \sum_{j=1}^{m_{i(\cdot)}} [y_{ij} - \{\bar{y}_{(\cdot)} + \hat{\beta}_{(\cdot)}(x_{i(\cdot)} - \bar{x}_{(\cdot)})\}]^2 \qquad (12)$$

(Step 2) In order to compile a control chart of error variances, the weighted estimator of the mean value of variances is determined by the following equation:

$$\bar{\hat{\sigma}}_i^2 = \frac{(n_{I1} - 2)\hat{\sigma}_1^2 + (n_{I2} - 2)\hat{\sigma}_2^2}{n_{I1} + n_{I2} - 4} \qquad (13)$$

which is also used as the center line in the control chart.

(Step 3) The upper confidence limit of the control chart will be

$$(\text{UCL})_{(\cdot)} = \frac{\widehat{\sigma}_i^2 \cdot \chi_{\phi(\cdot)}^2(\gamma/2)}{\phi_{(\cdot)}} \tag{14}$$

In this case, the lower confidence limit is not necessary.

(Step 4) If the residual square between the data and fitted equation $\widehat{\sigma}_{(\cdot)}^2$ is smaller than the upper confidence limit, i.e. if

$$\widehat{\sigma}_{(\cdot)}^2 < (\text{UCL})_{(\cdot)}$$

then the hypothesis for the test condition 2 will be accepted; otherwise it is rejected.

(3) *Condition 3: test of equality of the slope of regression equations.*

(Step 1) Calculate the weighted estimator of the mean of regression coefficient $\widehat{\beta}_{(\cdot)}$ as

$$\overline{\beta} = \frac{\displaystyle\sum_{i=1}^{l_1}(x_{i1} - \overline{x}_1)^2\widehat{\beta}_1 + \sum_{i=1}^{l_2}(x_{i2} - \overline{x}_2)^2\widehat{\beta}_2}{\displaystyle\sum_{i=1}^{l_1}(x_{i1} - \overline{x}_1)^2 + \sum_{i=1}^{l_2}(x_{i2} - \overline{x}_2)^2} \tag{15}$$

which is used for the center line of the control chart of $\widehat{\beta}_{(\cdot)}$.

(Step 2) The upper and lower limits of the control chart are given by

$$(\text{UCL})_{(\cdot)} = \overline{\beta} + t_{\phi(\cdot)}(\gamma/2) \cdot \sqrt{V_{e(\cdot)} \cdot T_{1(\cdot)}}$$
$$(\text{LCL})_{(\cdot)} = \overline{\beta} - t_{\phi(\cdot)}(\gamma/2) \cdot \sqrt{V_{e(\cdot)} \cdot T_{1(\cdot)}} \tag{16}$$

where

$$V_{e(\cdot)} = \frac{1}{n_{I(\cdot)} - 2}\sum_{i=1}^{l_{(\cdot)}}\sum_{j=1}^{m_{i(\cdot)}}\{y_{ij} - (\widehat{\alpha}_{(\cdot)} + \widehat{\beta}_{(\cdot)}x_{i(\cdot)})\}^2$$

$$T_{1(\cdot)} = \frac{1}{n_{I(\cdot)} \cdot S_{xx(\cdot)}}$$

(Step 3) When

$$(\text{LCL})_{(\cdot)} < \widehat{\beta}_{(\cdot)} < (\text{UCL})_{(\cdot)} \tag{17}$$

is satisfied, let $\overline{\beta}$, the new estimates, $\widehat{\beta}_*$, and the estimator of $\widehat{\sigma}^2$ be given by

$$\widehat{\sigma}^2 = \frac{(n_{I1} - n_{I2} - 4)\widehat{\sigma}^2 + \widehat{\sigma}_{**}^2}{n_{I1} - n_{I2} - 3} \tag{18}$$

where

$$\hat{\sigma}^2_{**} = \hat{\sigma}^2_{**}(\beta_*) = \frac{(\hat{\beta}_1 - \hat{\beta}_2)^2}{\dfrac{1}{\sum\limits_{i=1}^{l_1}(x_{i1} - \bar{x}_1)^2} + \dfrac{1}{\sum\limits_{i=1}^{l_2}(x_{i2} - \bar{x}_2)^2}}$$

(4) *Condition 4: test of equality of the intercepts of regression equation.*
(Step 1) Calculate the weighted estimator of the mean of regression coefficient $\hat{a}_{(\cdot)}$ as

$$\bar{\bar{a}} = \frac{\sum\limits_{i=1}^{l_1}(x_{i1} - \bar{x}_1)^2 \cdot \hat{a}_1 + \sum\limits_{i=1}^{l_2}(x_{i2} - \bar{x}_2)^2 \cdot \hat{a}_2}{\sum\limits_{i=1}^{l_1}(x_{i1} - \bar{x}_1)^2 + \sum\limits_{i=1}^{l_2}(x_{i2} - \bar{x})^2} \qquad (19)$$

Equation (19) is also used for the center line of the control chart of $\hat{a}_{(\cdot)}$.

(Step 2) The upper and lower confidence limits are

$$(\text{UCL})_{(\cdot)} = \bar{\bar{a}} + t_{\phi(\cdot)}(\gamma/2)\sqrt{V_{e(\cdot)} \cdot T_{2(\cdot)}} \qquad (20)$$

$$(\text{LCL})_{(\cdot)} = \bar{\bar{a}} - t_{\phi(\cdot)}(\gamma/2)\sqrt{V_{e(\cdot)} \cdot T_{2(\cdot)}}$$

where

$$T_{2(\cdot)} = \frac{1}{n_{I(\cdot)}} + \frac{\bar{x}^2_{(\cdot)}}{n_{I(\cdot)}S_{xx(\cdot)}} \qquad (21)$$

(Step 3) Plot the estimates of a on the control chart.

(Step 4) When $\hat{a}_{(\cdot)}$ is located between $(\text{UCL})_{(\cdot)}$ and $(\text{LCL})_{(\cdot)}$, the hypothesis is accepted and the new estimate of a is replaced by $\bar{\bar{a}}$, i.e.

$$\hat{a}_* = \bar{\bar{a}}$$

(5) *Condition 5: test of equality of mean values of fatigue limit.*

(Step 1) Calculate the weighted mean value of the estimator of fatigue strength, $\hat{E}_{(\cdot)}$, which is given by the Probit method as

$$\bar{\bar{E}} = \frac{n_{f1}\hat{E}_1 + n_{f2}\hat{E}_2}{n_{f1} + n_{f2}} \qquad (22)$$

Equation (22) is also used as the center line of the control chart of $E_{(\cdot)}$.

(Step 2) The control limits for the upper and lower sides are

$$(\text{UCL})_{(\cdot)} = \overline{\overline{E}} + t_{\phi(\cdot)}(\gamma/2)\frac{S_{(\cdot)}}{\sqrt{n_{I(\cdot)} - 1}}$$

$$(\text{LCL})_{(\cdot)} = \overline{\overline{E}} - t_{\phi(\cdot)}(\gamma/2)\frac{S_{(\cdot)}}{\sqrt{n_{I(\cdot)} - 1}}$$

(23)

where

$$S_{(\cdot)} = \sqrt{\frac{1}{n_{I(\cdot)} - 1}\sum_{i=1}^{l}(x_{i(\cdot)} - \widehat{E}_{(\cdot)})^2}$$

(24)

(Step 3) If the estimates $\widehat{E}_{(\cdot)}$ are located within the control line, the hypothesis that the means of the horizontal parts are the same is accepted.

(Step 4) Otherwise, when $\widehat{E}_{(\cdot)}$ is larger than $(\text{UCL})_{(\cdot)}$ or smaller than $(\text{LCL})_{(\cdot)}$, the data cannot be pooled.

(6) *Condition 6: test of equality of variances of horizontal part.*

(Step 1) Calculate the estimates of variance of fatigue strength as

$$\widehat{\sigma}_{h(\cdot)}^2 = \frac{1}{n_{I(\cdot)} - 1}\sum_{i=1}^{l}(x_{i(\cdot)} - \widehat{E}_{(\cdot)})^2$$

(25)

(Step 2) The weighted mean value of $\widehat{\sigma}_{h(\cdot)}^2$ is obtained by

$$\overline{\widehat{\sigma}}_h^2 = \frac{(n_{I1} - 1)\widehat{\sigma}_{h1}^2 + (n_{I2} - 1)\widehat{\sigma}_{h2}^2}{n_{I1} + n_{I2} - 2}$$

(26)

which is also used as the center line of the control chart.

(Step 3) The only upper confidence limit is calculated by

$$(\text{UCL})_{(\cdot)} = \frac{\overline{\widehat{\sigma}}_h^2 \cdot \chi_{\phi(\cdot)}^2(\gamma/2)}{n_{I(\cdot)} - 1}$$

(27)

(Step 4) If $\widehat{\sigma}_{h(\cdot)}^2$ is smaller than $(\text{UCL})_{(\cdot)}$, this condition is satisfied.

In the above conditions 2–6, the graphic techniques developed by the author [9] which utilize the control charts are useful for evaluation.

In the proposed method, two kinds of samples of fatigue data can be pooled when all the conditions are satisfied. The details of the steps in the tests of each condition and the mathematical derivations of the equations used in each condition are given elsewhere [9, 10].

Pooling Method for Multiple Samples of Fatigue Data

In order to apply the pooling method of two samples to the problem of pooling multiple fatigue data, an algorithm is developed, the calculation flow chart of which is shown in Fig. 3. The key points of this algorithm are: (1) to order the samples according to size $D_1, D_2, \ldots, D_{k_0}$; (2) to determine which samples satisfy conditions 1, 2, 3, 4, 5 or 6,

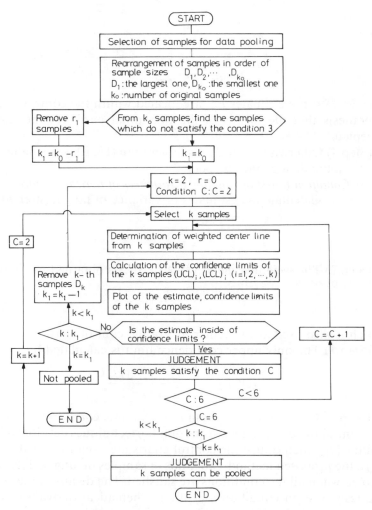

FIG. 3. Procedure of judgment for multiple samples.

sequentially. If it is found that a sample does not satisfy at least one condition, it should be eliminated. In the algorithm, statistical tests on conditions 2, 3, 4, 5 and 6 are repeated until all the data sets are checked as shown in Fig. 3.

In this algorithm for multiple samples, the following equations should be used to modify the details of the conditions 2–6.

Condition 2. The weighted estimate of the mean value of variances is

$$\bar{\bar{\sigma}}_i^2 = \frac{\sum\limits_{j=1}^{k} (n_{lj} - 2)\hat{\sigma}_j^2}{\sum\limits_{j=1}^{k} n_{lj} - 2k} \qquad (k = 2, 3, \ldots, k_1) \qquad (28)$$

instead of Eqn. (13).

Condition 3. Instead of Eqn. (15), $\bar{\beta}$ is given by

$$\bar{\bar{\beta}} = \frac{\sum\limits_{j=1}^{k} \sum\limits_{i=1}^{l_j} (x_{ij} - \bar{x}_j)^2 \hat{\beta}_j}{\sum\limits_{j=1}^{k} \sum\limits_{i=1}^{l_j} (x_{ij} - \bar{x}_j)^2} \qquad (k = 2, 3, \ldots, k_1) \qquad (29)$$

If $\hat{\beta}_{(\cdot)}$ is within the control limit, the new estimate of β is $\bar{\bar{\beta}}$. Then the estimate of σ^2 is represented by

$$\hat{\sigma}_*^2 = \frac{\left(\sum\limits_{j=1}^{k} n_{lj} - 2k\right)\bar{\bar{\sigma}}_i^2 + \hat{\sigma}_{**}^2}{\sum\limits_{j=1}^{k} n_{lj} - 2k + 1} \qquad (k = 2, 3, \ldots, k_1) \qquad (30)$$

where

$$\hat{\sigma}_{**}^2 = \hat{\sigma}_{**(\bar{\beta})}^2 = \frac{\sum\limits_{i=1}^{k-1} \sum\limits_{j=i+1}^{k} (\hat{\beta}_i - \hat{\beta}_j)^2}{\sum\limits_{j=1}^{k} \left\{ \dfrac{1}{\sum\limits_{j=1}^{k} (x_{ij} - \bar{x})^2} \right\}} \qquad (k = 2, 3, \ldots, k_1)$$

Condition 4. Instead of Eqn. (19), $\overline{\overline{\alpha}}$ is calculated by the equation

$$\overline{\overline{\alpha}} = \frac{\displaystyle\sum_{j=1}^{k}\sum_{i=1}^{l_j}(x_{ij} - \overline{x}_j)^2 \cdot \widehat{\alpha}_j}{\displaystyle\sum_{j=1}^{k}\sum_{i=1}^{l_j}(x_{ij} - \overline{x}_j)^2} \qquad (k = 2, 3, \ldots, k_1) \qquad (31)$$

Condition 5. $\overline{\overline{E}}$ in Eqn. (22) is modified by

$$\overline{\overline{E}} = \frac{\displaystyle\sum_{j=1}^{k} n_{lj} \cdot \widehat{E}_j}{\displaystyle\sum_{j=1}^{k} n_{lj}} \qquad (k = 2, 3, \ldots, k_1) \qquad (32)$$

Condition 6. Instead of Eqn. (25), the equation

$$\overline{\overline{\sigma}}_h^2 = \frac{\displaystyle\sum_{j=1}^{k} (n_{lj} - 1)\widehat{\sigma}_{kj}^2}{\displaystyle\sum_{j=1}^{k} n_{lj} - k} \qquad (k = 2, 3, \ldots, k_1) \qquad (33)$$

is used for the estimator of the weighted mean value of variances of the fatigue limit.

APPLICATION OF FATIGUE DATA BASE TO RELIABILITY-BASED DESIGN

Information Table from Data Base

In consideration of this idea of data pooling, now let us run the practical data base on the fatigue strength of metallic materials developed by the Japanese Society of Materials Science [3], in order to utilize reliability-based design. By sorting the data base with respect to the materials JIS A2017 (aluminium alloy) and JIS S45C (45% carbon steel), Tables 1 and 2 were obtained. In these tables, the rows show the various kinds of shapes of the specimens, and the columns show the types of cyclic load conditions to which the specimens were subjected.

TABLE 1
Information table of JIS A2017 (*P-N* data)

NUMBER OF BLOCK/DATA (PN DATA)

	AX	RB	RC	B I	BO	BB	TW
SM	0 / 0	0 / 0	0 / 0	0 / 0	0 / 0	0 / 0	0 / 0
RD	0 / 0	0 / 0	0 / 0	0 / 0	2 / 160	0 / 0	0 / 0
NT	0 / 0	0 / 0	0 / 0	0 / 0	0 / 0	0 / 0	0 / 0
FL	0 / 0	0 / 0	0 / 0	0 / 0	0 / 0	0 / 0	0 / 0
SF	0 / 0	0 / 0	0 / 0	0 / 0	0 / 0	0 / 0	0 / 0
PF	0 / 0	0 / 0	0 / 0	0 / 0	0 / 0	0 / 0	0 / 0

Symbol of row
SM Smooth specimen
RD Hourglass type specimen
NT Notched specimen
FL Specimen with fillet
SF Shrinkage fit specimen
PF Press-fitted specimen

Symbol of column
AX Axial loading
RB Uniform rotating bending
RC Cantilever rotating bending
BI In-plane bending of plate
BO Ex-plane bending of plate
BB Bending of round bar
TW Torsion

TABLE 2
Information table of JIS S45C (*S-N* data)

NUMBER OF BLOCK/DATA (SN DATA)

	AX	RB	RC	B I	BO	BB	TW
SM	2 / 43	51 / 380	0 / 0	6 / 38	0 / 0	0 / 0	3 / 45
RD	5 / 49	7 / 115	1 / 10	0 / 0	0 / 0	0 / 0	2 / 32
NT	12 / 136	20 / 168	4 / 63	0 / 0	0 / 0	0 / 0	0 / 0
FL	0 / 0	2 / 20	0 / 0	0 / 0	0 / 0	0 / 0	0 / 0
SF	0 / 0	0 / 0	0 / 0	0 / 0	0 / 0	0 / 0	0 / 0
PF	0 / 0	0 / 0	0 / 0	0 / 0	0 / 0	0 / 0	0 / 0

For key to symbols, see Table 1.

In the data base, there are three kinds of fatigue data: $S-N$, $P-N$ and $S-T$. While $S-N$ data are fatigue test data obtained by ordinary testing methods to obtain a conventional $S-N$ relation, $P-N$ data are fatigue test data obtained by using a large number of specimens with the intention of statistical analysis of the distribution of fatigue lives. In addition, $S-T$ data are fatigue data to obtain the fatigue limit by the up-and-down method. Tables 1 and 2 show how much $P-N$ and $S-N$ data were obtained in this data base for JIS A2017 and JIS S45C materials under each fatigue test condition. In Table 2 for JIS S45C, it is seen that combination of notched specimen and rotary bending load (NT-RB) gives 20 blocks with a total of 168 test data.

Data Pooling from Fatigue Data Base

Two kinds of fatigue data were selected as candidates for pooling from A2017-RD-BO (material = aluminium alloy; specimen = hourglass type; load = ex-plane bending of plate), code series 169220 and 169221. Table 3 shows a summary of the results for the test of pooling two kinds of A2017-RD-BO fatigue data whose $S-N$ curves are determined by bi-linear curve fitting as shown in Fig. 4. In this table, it is seen that the four conditions 1-4 for inclined parts are satisfied and that the other two conditions (5 and 6) for horizontal parts are also satisfied.

TABLE 3
Judgment results of pooling candidates for JIS A2017-RD-BO, 189-220, 221

	Sample 1	Sample 2		Condition 4	
n_I	80	80	$\bar{\alpha}$	17.9	
l	4	4	(UCL)	22.3	22.7
Condition 1			$\hat{\alpha}$	19.9	16.1
F_o	1.96	2.28	(LCL)	13.1	13.3
$F(l-2,n-l:Y)$	5.01	4.27	Result	accept	
Result	accept			$\bar{\alpha}_* = 17.9$	
Condition 2				Sample 1	Sample 2
(UCL)	0.225	0.227	n_I	57	80
$\hat{\sigma}_I{}^2$	0.131	0.161	Condition 5		
Result	accept		$\bar{\hat{\epsilon}}$	2.239	
	$\bar{\sigma}_I^2 = 0.146$		(UCL)	2.276	2.276
Condition 3			$\hat{\epsilon}$	2.237	2.240
$\bar{\beta}$	-4.96		(LCL)	2.201	2.201
(UCL)	-3.16	-2.98	Result	accept	
$\hat{\beta}$	-5.66	-4.27	Condition 6		
(LCL)	-6.77	-6.94	$\bar{\hat{\sigma}}_h{}^2$	0.0159	
Result	accept		(UCL)	0.0233	0.0233
	$\hat{\beta}_* = -4.96, \hat{\sigma}_*^2 = 0.144$		$\hat{\sigma}_h{}^2$	0.0161	0.0155
	$\bar{\sigma}_{**}^2 = 0.004$		Result	accept	

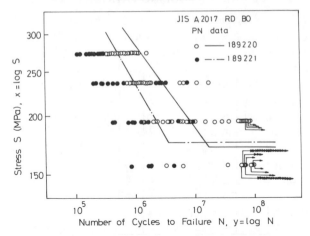

FIG. 4. *S–N* plot of fatigue data for JIS A2017-RD-BO.

Figure 5 shows the control charts that are utilized to judge whether the samples satisfy the conditions 2–6 or not. Thus, it is determined that two samples are pooled because all the hypotheses are accepted. Figure 6 shows *P–S–N* curves drawn using data pooled.

Another trial is performed for multiple samples of S45C-NT-RC (material = 45% carbon steel for structural element; specimen = notched type; load = cantilever rotating bending) in the data base on fatigue strength of metallic materials by the Japanese Society of Materials Science. There are four kinds of candidates for pooling whose code series are 189070, 189073, 120006, 189071; the total number of specimens is 63. The relationships between applied stress and number of cycles to failure of the original data for S45C-NT-RC are illustrated in Fig. 7, where four different samples are plotted with respective marks. As a result of the pooling analysis, three samples among four candidates are accepted for pooling. On the other hand, one sample of code series 120006 is rejected, since in this case condition 5, which means judgment of equality of mean of fatigue limit, is not satisfied. Figure 8 shows the *P–S–N* (Probability–Stress–Number) curves derived from the samples pooled. This figure enables the designer to take into account the probabilistic consideration of fatigue properties of JIS S45C-NT-RC, because this figure concentrates much information from the three kinds of fatigue test results.

In Fig. 7 the fatigue limit of sample (4) whose code series number is 189071 is not valid, because these data cover only the range 10^4–10^5

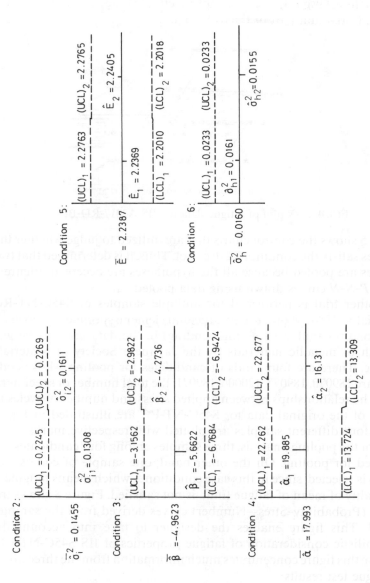

FIG. 5. Control chart of parameters from JIS A2017-RD-BO.

cycles. Therefore the candidates for pooling with no fatigue limit or without long fatigue life data are not effective to judge conditions 5 and 6 for the horizontal part of the *S–N* data.

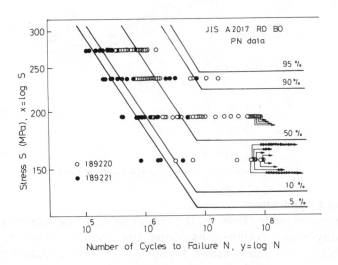

FIG. 6. *P–S–N* curves for JIS A2017-RD-BO.

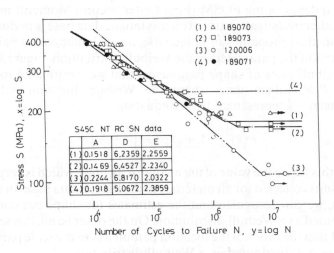

FIG. 7. *S–N* plots for JIS S45C-NT-RC.

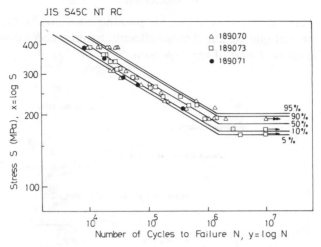

FIG. 8. *P–S–N* curves for JIS S45C-NT-RC.

Advanced Applications

The results pooled from the data base enable the designer to perform reliability-based design considering the factors of safety, failure probability, and safety index. Figure 9 is a schematization of these objectives, where the relationship between the central factor of safety and the probability of failure by non-parametric (Weibull) methods was previously developed by the author [11]. This figure also shows the safety indexes in the FOSM (First Order Second Moment) method.

Another investigation based on this fatigue data base is performed on the statistical properties of fatigue life. Special attention is paid to the estimates of the parameters of the Weibull distribution. Figure 10 shows the Weibull plots of shape parameters that were estimated for all JIS S45C *P–N* fatigue data where the Weibull function with three parameters is represented by the equation

$$F(x) = 1 - \exp\left\{-\left(\frac{x - \gamma}{\beta}\right)^{\alpha}\right\} \tag{34}$$

From this figure the value of the median $\tilde{\alpha}$ is 1·696, which is very close to the results collected for all materials in the fatigue data base. In addition to this, we can recognize that the estimates for shape parameters are distributed as a Weibull distribution. On the other hand, it is seen from Fig. 11 that the ratios of the location parameters to the scale parameters, α/β, are not distributed as a Weibull distribution.

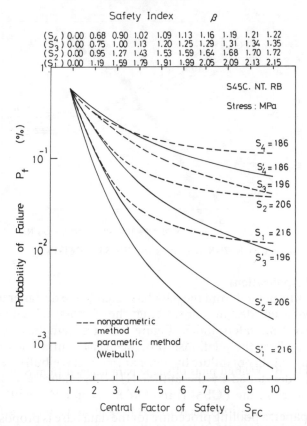

FIG. 9. Relation between probability of failure, central factor of safety, and safety index corresponding to each stress level.

CONCLUDING REMARKS

The role of pooling of fatigue data is the key to making the fatigue data base useful in utilizing the aggregate knowledge for the future design of structures. Therefore it is especially necessary to develop the procedure of pooling multiple samples. It is suggested that the method proposed by the author satisfies the above needs. The practical application examples with use of the data base on fatigue strength of metallic materials collected by the Japanese Society of Materials Science would verify that the method is useful.

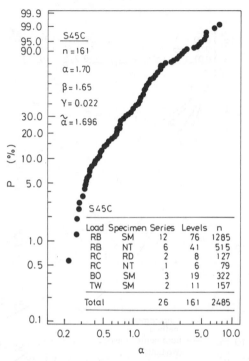

FIG. 10. Weibull plots of shape parameters α from S45C.

In this paper, a pooling procedure for the data base is proposed on the assumption that $y = \log N$ and $x = \log S$. In the practical use of S-N data, however, the stress amplitude, S, is often plotted on a linear scale. It is easily seen that the proposed pooling method is applicable in such a case with the transformations of $x = S$ instead of $x = \log S$.

ACKNOWLEDGEMENTS

This research owes a great deal to the working group on the Data Book on Fatigue Strength of Metallic Materials, the Committee on Fatigue and the Committee on Reliability Engineering of the Japanese Society of Materials Science. The author expresses great thanks to Prof. Dr. T. Tanaka and Dr. T. Sakai, both of Ritsumeikan University, and to the members of these committees. The author is also very grateful to

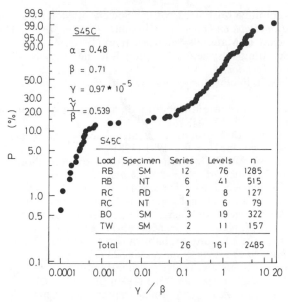

FIG. 11. Weibull plots of the ratios γ/β from S45C.

Prof. Dr. H. Ohta, of Osaka Prefecture University, and to Dr. S. Nishijima, of the National Research Institute of Metallurgy, for their helpful suggestions. Additional thanks are due to T. Umehara, Y. Tsubooka and T. Murakami, students at the Kanazawa Institute of Technology, for their help in developing the necessary computer programming. Finally, it is noted that this research was partly supported by the Science Research Fund of the Ministry of Education, Science and Culture of Japan.

REFERENCES

[1]   J. A. Graham, *The Metal Properties Data Storage and Retrieval System,* ASME MPC-14, 23 (1980).

[2]   C. L. Blackburn, O. O. Storoali and R. E. Fulton, The role and application of data base management in integrated computer-aided design. *J. Aircraft,* **20**, 717 (1983).

[3]   The Committee on Fatigue and the Committee on Reliability Engineering of the Society of Materials Science, Japan (Eds.), *Data Book on Fatigue Strength of Metallic Materials,* Vols. 1, 2, 3, Society of Materials Science, Japan (1982).

[4] The Committee on Fatigue of the Society of Materials Science, Japan (Eds.), *Data Book on Fatigue Crack Growth Rates of Metallic Materials*, Vols. 1, 2, Society of Materials Science, Japan (1982).

[5] E. G. Schilling, A systematic approach to the analysis of means, Part I, *J. Quality Technol.*, **5**, 93 (1983).

[6] L. S. Nelson, Factors for the analysis of means, *J. Quality Technol.*, **6**, 175 (1974).

[7] H. Ohta, A procedure for pooling data by analysis of means, *J. Quality Technol.*, **13**, 115 (1981).

[8] Japan Society of Mechanical Engineers, *Standard Method of Statistical Fatigue Testing*, JSME S 002-1981 (1981).

[9] H. Nakayasu and T. Umehara, Study on pooling method of fatigue data for reliability design, *Proc. 27th Japan Congr. Mater. Res.*, Society of Materials Science, Japan, 111 (1984).

[10] H. Nakayasu and T. Umehara, Study on pooling method of fatigue data, *Trans. Japan Soc. Mech. Engrs., Ser. A*, **51**, 2076 (1985) (in Japanese).

[11] Y. Murotsu, H. Nakayasu, K. Mori and S. Kase, New method of relating safety factor to failure probability in structural design, *Advances in Reliability and Stress Analysis*, ASME, 23 (1979).

[12] S. Nishijima, Statistical analysis of small sample from fatigue data, *Trans. Japan Soc. Mech. Engrs.*, **46**, 1303 (1980) (in Japanese).

[13] S. Nishijima and A. Ishii, Parameter representation of S-N data by the bi-linear curve fitting, *J. Soc. Mater. Sci., Japan*, **34**, 340 (1985) (in Japanese).

Stochastic Fatigue Crack Growth Model and Its Wide Applicability in Reliability-based Design

HIROSHI ISHIKAWA

Department of Information Science, Kagawa University, 2-1 Saiwai-cho, Takamatsu 760, Japan

AKIRA TSURUI

Department of Applied Mathematics and Physics, Kyoto University, Yoshida-honmachi, Sakyo-ku, Kyoto 606, Japan

and

HITOSHI KIMURA

Kagawa University, 1-1 Saiwai-cho, Takamatsu 760, Japan

ABSTRACT
This paper proposes a stochastic fatigue crack growth model in consideration of spatially distributed random propagation resistance. First, utilizing the Markov approximation method, a generalized Fokker–Planck equation to describe the temporal variation of the crack length distribution is set forth and solved analytically, both for locally stationary random loading and for random propagation resistance. The solution of this equation yields a stochastic crack growth model reflecting the engineering reality of random propagation resistance due to material inhomogeneity. Next, the effect of uncertainty associated with initial flaw size is theoretically evaluated on the temporal variation of the crack size distribution with the aid of the proposed stochastic growth model. Finally, the relationship of the reliability degradation of the material to the temporal variation of the crack size distribution is discussed under the assumption that a structural component may contain some initial flaws, the number of which follows a Poisson distribution.

INTRODUCTION

The damage-tolerant design philosophy has been widely applied to those structures whose failure might cause a large loss of human life, including aircraft, space and deep sea structures. Any structure can fail due to fatigue resulting from the repetition of randomly applied loads. However, fatigue life or strength of the material usually shows a wide scatter even under the same loading conditions. Therefore, a good understanding of the statistical aspects of fatigue is unquestionably of great importance in reliability-based design of structural components, and there has been a variety of research to date dealing with

45

uncertainty in fatigue life or strength of the material under random loading [1-26].

In this respect, the present paper is concerned with the clarification of the effect of spatially distributed random crack propagation resistance, due to lack of material homogeneity, on the variation of the crack length distribution and, consequently, on the crack propagation life distribution, since the actual materials used in structural components will inevitably involve, to a degree, non-homogeneous micro-structures which will tend to cause spatial randomness of the propagation resistance.

Random crack propagation resistances at two different locations sufficiently far apart can be regarded as statistically independent of each other. Together with the consideration of slow crack growth, therefore, the term of correlation would be deemed sufficiently short in comparison with the number of cycles of interest to investigate the crack increment. This justifies the use of the Markov approximation method [27-29].

In view of the above, first a generalized Fokker–Planck equation was derived to describe the temporal variation of the crack length distribution for locally stationary random loading and locally stationary random propagation resistance. Then the equation was solved analytically to form a useful stochastic crack growth model.

Secondly, the effect of uncertainty associated with the initial flaw size has been theoretically evaluated on the temporal variation of the crack size distribution with the aid of this stochastic crack growth model. Further, discussion is centered on how the reliability of an arbitrary structural component varies with the number of stress cycles under the assumption that the component might have some initial flaws, the number of which follows a Poisson distribution.

A GENERALIZED FOKKER–PLANCK EQUATION AND ITS SOLUTION

The Paris–Erdogan crack growth law can be expressed as

$$dX/dn = \varepsilon(\Delta K/\Delta K_0)^m \tag{1}$$

where $X = l/l_0$ represents the non-dimensional crack length after n cycles of loading, ΔK and ΔK_0 the stress intensity factor range associated with applied load and its standardization factor, and ε and m are material constants. In the present study, the following relationship is assumed for simplicity:

$$\Delta K / \Delta K_0 = S\sqrt{\pi l}/(s_0\sqrt{\pi l_0}) = Z\sqrt{X} \tag{2}$$

in which $Z = S/s_0$ is the appropriately normalized stress amplitude. Suppose that the following stochastic differential equation holds, as a natural extension of Eqns. (1) and (2), in the case of random propagation resistance and/or random applied load:

$$dX/dn = \varepsilon C_n Z_n^{2(\lambda + 1)} X^{\lambda + 1} \tag{3}$$

where C_n represents the non-dimensional locally stationary random propagation resistance, such that $E[C_n] = 1$ and $Z_n = S_n/s_0$, the non-dimensional locally stationary random stress amplitude. The constant λ of the form

$$\lambda = m/2 - 1 \tag{4}$$

is introduced in place of m in the above equation for later convenience. Generally speaking, C_n can be chosen to be of the order of 1. With C_n thus chosen, ε takes on a sufficiently small positive constant value and becomes a non-negative material constant.

However, in dealing with this problem, careful attention must be paid to the possibility where the crack length tends to infinity after a limited number of stressings. Thus, the death point $\{\infty\}$, which corresponds to the state of infinite crack length, needs to be involved in the state space.*

*It is undoubtedly essential to introduce the idea of the 'death point' in studying the stochastic nature of fatigue crack growth. The reason is apparent from the following. The Paris–Erdogan deterministic law of crack growth under constant stress amplitude Z_0 can be given in a normalized form, as in Eqn. (3), as

$$dx/dn = \varepsilon Z_0^{2(\lambda + 1)} x^{\lambda + 1} \tag{f.1}$$

where x represents normalized crack length after n cycles of stressing, ε is a constant and λ a kind of power exponent assuming $\lambda \geqslant 0$. Solving this differential equation with the initial condition $x = x_i$ at $n = 0$, we get

$$x_i^{-\lambda} - x^{-\lambda} = \lambda \varepsilon Z_0^{2(\lambda + 1)} n \tag{f.2}$$

As is clearly seen from Eqn. (f.2), the crack length x tends monotonically to infinity with increasing value of n. For values of n not less than $n_0 = x_i^{-\lambda}(\lambda \varepsilon Z_0^{2(\lambda + 1)})^{-1}$, x no longer assumes a real number. This discussion could be regarded as the case of a sample function of the stochastic fatigue crack growth process when Z_0 is replaced by the random stress amplitude Z_n and x by a random variable of crack length X. However, there are some other sample functions of which the values still remain less than infinity even if n becomes greater than the aforementioned value of n_0. This is really the case when dealing with stochastic processes, and hence the insufficiency arises as long as we consider the state space as one-dimensional Euclidian space.

Let $X(n)$ be the non-dimensional crack length after n cycles of stressing, let

$$\Pr[X(n) \in \{\infty\}] = P_{\infty}(n) \tag{5}$$

be the probability that $X(n)$ lies in the death point, and let

$$\Pr[X(n) \leq x \,|\, X(0) = x_i] = W(x, n \,|\, x_i) \tag{6}$$

be the probability that the crack of initial size x_i grows to a finite value less than or equal to x after n cycles of stressing. Then it is evident that the following relationship holds:

$$\lim_{x \to \infty} W(x, n \,|\, x_i) + P_{\infty}(n) = 1 \tag{7}$$

Suppose that the function $W(x, n \,|\, x_i)$ is differentiable with respect to x and its derivative is $w(x, n \,|\, x_i)$.

By use of the Markov approximation method discussed in reference [27], a generalized Fokker–Planck equation to describe the temporal variation of the crack length distribution $w(x, n \,|\, x_i)$ can be derived as

$$\frac{\partial w}{\partial n} = -\beta(n) \frac{\partial}{\partial x} \{x^{\lambda + 1} w\} - \frac{\lambda + 1}{2} \gamma(n) \frac{\partial}{\partial x} \{x^{2\lambda + 1} w\}$$

$$+ \frac{1}{2} \gamma(n) \frac{\partial^2}{\partial x^2} \{x^{2(\lambda + 1)} w\} \tag{8}$$

where $\beta(n)$ and $\gamma(n)$ are defined in the following forms:

$$\beta(n) = \varepsilon \mathrm{E}[C_n Z_n^{2(\lambda + 1)}] \tag{9}$$

$$\gamma(n) = 2\varepsilon^2 \int_{-\infty}^{0} \{\mathrm{E}[C_n Z_n^{2(\lambda + 1)} C_{n + n'} Z_{n + n'}^{2(\lambda + 1)}]$$

$$- \mathrm{E}[C_n Z_n^{2(\lambda + 1)}] \mathrm{E}[C_{n + n'} Z_{n + n'}^{2(\lambda + 1)}]\} \, dn' \tag{10}$$

in which $\mathrm{E}[\,\cdot\,]$ denotes an operator to involve the expectation of random variates.

After some mathematical manipulation, the solution of the Fokker–Planck equation (8) that satisfies

$$\lim_{x \to 0} w(x, n \,|\, x_i) = \delta(x - x_i)$$

can be obtained as follows:

$$w(x, n \,|\, x_i) = \frac{1}{x^{\lambda + 1} \sqrt{2\pi \int_0^n \gamma(n')\,dn'}}$$

$$\times \exp\left[-\frac{\left\{x_i^{-\lambda} - x^{-\lambda} - \lambda \int_0^n \beta(n')\,dn'\right\}^2}{2\lambda^2 \int_0^n \gamma(n')\,dn'}\right] \qquad (11)$$

CRACK LENGTH DISTRIBUTION

A simple but important case is considered here where applied load is deterministic with a constant amplitude of Z_0. It is reasonable to assume uniformly distributed spatial random non-homogeneity of the material. Under these conditions, the mean crack propagation resistance $E[C_n]$ becomes independent of n. Hence, $\beta(n)$ in Eqn. (9) also becomes independent of n as in the following:

$$\beta(n) = \varepsilon Z_0^{2(\lambda + 1)} E[C_n] \qquad (12)$$

Further, from the spatial viewpoint, the propagation resistance can be considered to have a significant correlation only within a certain finite distance referred here to the correlation distance. This correlation distance can be considered independent of the location for homogeneous materials. On the other hand, the correlation of the crack propagation process needs to be discussed from the temporal viewpoint. Thus we introduce the correlation time which is defined as the time difference within which the correlation takes a significant value. Although the correlation distance can be considered constant at any location, this correlation time will become gradually shorter, since the passage time of the correlation distance is comparatively large when the crack propagation rate is small, whereas it becomes shorter and shorter when the crack propagation rate increases. In this respect, it seems appropriate to assume that the correlation time is inversely proportional to the average propagation rate, dx/dn, at the location concerned. Here, $\gamma(n)$ in Eqn. (10), which is proportional to the correlation time, can be given finally as

$$\gamma(n) = \varepsilon^2 Z_0^{4(\lambda + 1)} \{E[C_n^2] - (E[C_n])^2\} \xi_0/(dx/dn)$$

$$= \begin{cases} \varepsilon Z_0^{2(\lambda + 1)} \xi_0 \dfrac{E[C_n^2] - (E[C_n])^2}{E[C_n]} \{x_i^{-\lambda} - \lambda\beta(n)n\}^{(\lambda + 1)/\lambda} \\ \qquad\qquad\qquad\qquad\text{(for } x_i^{-\lambda} > \lambda\beta(n)n) \qquad (13) \\ 0 \qquad\qquad\qquad\qquad\qquad \text{(otherwise)} \end{cases}$$

where ξ_0 is introduced as a proportional constant. With Eqns. (11), (12) and (13), the probability, $W(x, n | x_i)$, defined by Eqn. (6) can be given as

$$W(x, n | x_i) = \int_0^x w(x, n | x_i) \, dx$$

$$= \Phi\left[\frac{x_i^{-\lambda} - x^{-\lambda} - \lambda\beta(n)n}{\lambda\sqrt{\dfrac{\alpha}{2\lambda + 1}\{x_i^{-(2\lambda + 1)} - (\text{Max}[0, x_i^{-\lambda} - \lambda\beta(n)n])^{(2\lambda + 1)/\lambda}\}}}\right] \quad (14)$$

where $\Phi[\cdot]$ is the standardized normal distribution function and where the parameter α, defined as

$$\alpha = \xi_0\{E[C_n^2] - (E[C_n])^2\}/(E[C_n])^2 \quad (15)$$

is the only parameter left unknown.

The above discussion is limited to the case where an initial flaw is of a deterministic size. As for practical applications, however, there might be many different types of cases, such as the case with no flaws, with many flaws, and so forth. Even for the case with a single flaw, it is often the case that its precise size cannot be determined but can only be presumed. Hence, the initial flaw size, X_i, has to be treated as a random variable with a distribution function $G(x_i)$.

In the present study, it is assumed that the distribution is of a two-parameter Weibull type [11] such that

$$G(x_i) = 1 - \exp\left\{-\left(\frac{x_i}{x_0}\right)^\nu\right\} \quad (16)$$

where ν is the shape and x_0 is the scale parameter of a two-parameter Weibull distribution. Then the crack length distribution, $W(x, n)$, after n cycles of stressing is related to the conditional distribution function $W(x, n | x_i)$ as

$$W(x, n) = \int_0^\infty W(x, n | x_i) \, dG(x_i) \quad (17)$$

By substituting Eqns. (14) and (16) into Eqn. (17) and by introducing the new variables

$$\left.\begin{array}{l} \tau = \lambda\beta(n)x_0^\lambda n \\ \xi = x/x_0 \\ \xi_i = x_i/x_0 \end{array}\right\} \tag{18}$$

and a new parameter

$$a = \sqrt{(2\lambda + 1)x_0/\alpha}/\lambda \tag{19}$$

the crack length distribution $W(x, n)$ reduces to

$$W(\xi, \tau) = v \int_0^\infty \Phi\left[\frac{a(\xi_i^{-\lambda} - \xi^{-\lambda} - \tau)}{\sqrt{\xi_i^{-(2\lambda + 1)} - (\text{Max}[0, \xi_i^{-\lambda} - \tau])^{(2\lambda + 1)/\lambda}}}\right]$$

$$\times \; \xi_i^{v - 1} \exp[-\xi_i^v] \, d\xi_i \tag{20}$$

It should be noted here that, if ξ is fixed to ξ_c in Eqn. (20), then $W(\xi_c, \tau)$ approximates the probability that the crack length is not greater than ξ_c at time τ, i.e. the probability that the crack propagation life is not less than τ. Consequently, $1 - W(\xi_c, \tau)$ becomes the life distribution for the crack propagating up to a critical length ξ_c [29].

Further, it is of interest to consider some special cases. First let us consider the case where $a \to \infty$. This corresponds to the case where there is no randomness in the crack propagation resistance. In this case we have the following relationship:

$$\lim_{a \to \infty} \Phi\left[\frac{a(\xi_i^{-\lambda} - \xi^{-\lambda} - \tau)}{\sqrt{\xi_i^{-(2\lambda + 1)} - (\text{Max}[0, \xi_i^{-\lambda} - \tau])^{(2\lambda + 1)/\lambda}}}\right]$$

$$= Y(\xi_i^{-\lambda} - \xi^{-\lambda} - \tau) \tag{21}$$

in which $Y(u)$ represents the unit step function defined by

$$Y(u) = \begin{cases} 1 & \text{for } u \geqslant 0 \\ 0 & \text{otherwise} \end{cases} \tag{22}$$

Hence, Eqn. (20) reduces to

$$W(\xi, \tau) = v \int_0^\infty Y(\xi_i^{-\lambda} - \xi^{-\lambda} - \tau)\xi_i^{v - 1} \exp[-\xi_i^v] \, d\xi_i$$

$$= 1 - \exp[-(\xi^{-\lambda} + \tau)^{-v/\lambda}] \tag{23}$$

Next is the case where $v \to \infty$. In this case, Eqn. (20) reduces to

$$W(\xi, \tau) = \Phi \left[\frac{a(1 - \xi^{-\lambda} - \tau)}{\sqrt{1 - (\text{Max}[0, 1 - \tau])^{(2\lambda + 1)/\lambda}}} \right] \quad (24)$$

and this corresponds to the case of deterministic initial flaw size. In this equation, if we consider the limiting case of $a \to \infty$, we have

$$W(\xi, \tau) = Y(1 - \xi^{-\lambda} - \tau) \quad (25)$$

This is precisely the case where every quantity is deterministic.

Figure 1 exemplifies the variation in the crack length distribution, $W(\xi, \tau)$, as a function of ξ with fixed values of $\lambda = 0.5$ and $v = 2.0$ for parameter values $a = 3.0, 5.0, 10.0$ and ∞. In each figure, the dotted lines indicate the levels to which the distribution, $W(\xi, \tau)$, tends when ξ approaches infinity. The difference between these levels and unity gives the probability that the state lies in the death point $\{\infty\}$. Since the abscissa is plotted on a logarithmic scale, it is clearly seen how fast longer cracks grow in comparison to shorter ones. In other words, at earlier stages the crack length distribution skews to the right (i.e. to the greater crack-length side) and subsequently it flows more and more into the death point $\{\infty\}$. Equations (15) and (19) show that the uncertainty in the propagation resistance decreases gradually as the value of the parameter a increases, and hence it can be anticipated that the crack length distribution takes a smoother shape. However, since little difference is observed among distributions for $a \geqslant 10.0$, as in Figs. 1(c) and 1(d), the uncertainty factor associated with randomness of the propagation resistance becomes less important for material with $a \geqslant 10.0$.

RELIABILITY DEGRADATION OF STRUCTURAL COMPONENTS

Equation (20) in the preceding section indicates the temporal variation of the crack length distribution when the initial flaw size follows a two-parameter Weibull distribution. Hence the result can be utilized to clarify the reliability degradation of a structural component which might involve plural initial flaws.

As noted above, the quantity $1 - W(\xi_c, \tau)$ approximates the crack propagation life distribution of a structural component with a single initial flaw that grows to a specified size of ξ_c $(= x_c/x_0)$. This can be understood easily because the probability that the initial flaw takes on a

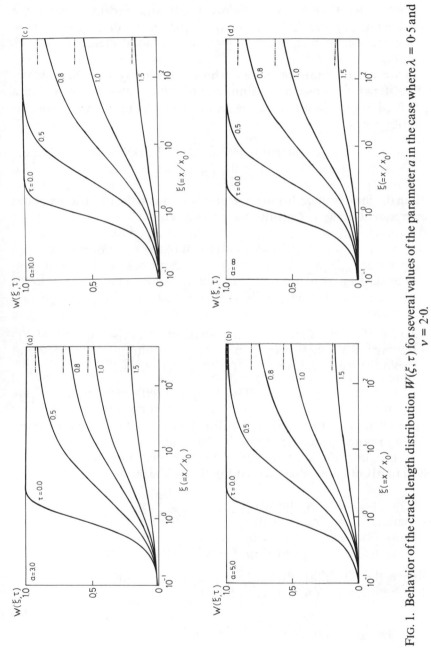

FIG. 1. Behavior of the crack length distribution $W(\xi, \tau)$ for several values of the parameter a in the case where $\lambda = 0.5$ and $\nu = 2.0$.

value greater than ξ_c is negligibly small and because, under the Markovian approximation method as applied in the present study, crack movement is limited to growth and no decrease in size is possible.

Now, a structural component chosen randomly may have many initial flaws. Suppose that the number of initial flaws, J, initiated in a production process is of a stochastic nature with the following Poisson distribution:

$$\left. \begin{array}{l} \Pr[J = j] = \dfrac{\mu^j \exp[-\mu]}{j!} \\[2mm] \mu = E[J] \end{array} \right\} \tag{26}$$

Then the joint probability of the events $\{J = j\}$ and the life of a structural component being not greater than τ is given by

$$\Pr[J = j]H^{(j)}(\tau) = \begin{cases} \dfrac{\mu^j \exp[-\mu]}{j!} \{1 - W(\xi_c, \tau)^j\} & \text{for } \tau < \infty \\[3mm] \dfrac{\mu^j \exp[-\mu]}{j!} & \text{for } \tau = \infty \end{cases} \tag{27}$$

where each initial flaw grows independently. In this equation, $H^{(j)}(\tau)$ represents the crack propagation life distribution of a structural component which has exactly j ($j \neq 0$) initial flaws:

$$H^{(j)}(\tau) = \begin{cases} 1 - W(\xi_c, \tau)^j & \text{for } \tau < \infty \\ 1 & \text{for } \tau = \infty \end{cases} \tag{28}$$

since $H^{(j)}(\tau)$ can be taken as the distribution of the minimum value among j random samples from the distribution $1 - W(\xi_c, \tau)$.

It is straightforward to derive the life distribution $H(\tau)$ of the structural component based on the notion of the marginal distribution:

$$\begin{aligned} H(\tau) &= \sum_{j=0}^{\infty} \frac{\mu^j \exp[-\mu]}{j!} \{1 - W(\xi_c, \tau)^j\} \\ &= 1 - \exp[-\mu\{1 - W(\xi_c, \tau)\}] \qquad \text{for } \tau < \infty \end{aligned} \tag{29}$$

Hence, the reliability, $R(\tau) = 1 - H(\tau)$, at time τ, can be related to the distribution function $W(\xi, \tau)$ as follows:

$$\frac{1}{\mu} \ln \frac{1}{R(\tau)} = 1 - W(\xi_c, \tau) \qquad \text{for } \tau < \infty \tag{30}$$

This relationship is illustrated in Figs. 2 and 3 for several values of the parameters a and v. It should be noted here that the parameter a gives a reciprocal measure of the randomness in the propagation resistance, i.e. the greater randomness the less the value of a. Little information has been available thus far for estimating the value of a for commercial materials other than that for a particular kind of high tensile strength material [30] where the estimated value of a is about 5·5 where $x_c = 0·3$ mm and $\lambda = 0·5$.

For those materials where $a \geqslant 10·0$, the effect of the randomness of the propagation resistance may not appear clearly. In the higher

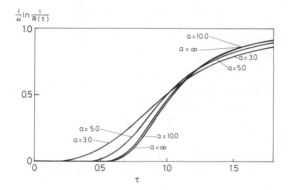

FIG. 2. Variation of the reliability $R(\tau)$ as a function of τ for several values of the parameter a ($\lambda = 0·5$, $v = 2·0$, $\xi_c = 256$).

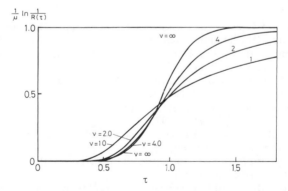

FIG. 3. Variation of the reliability $R(\tau)$ as a function of τ for several values of the parameter v ($\lambda = 0·5$, $a = 5·0$, $\xi_c = 256$).

reliability region, if x_0 assumes the same value, the difference in the value of the shape parameter does not greatly influence the reliability degradation, for $v \geqslant 2 \cdot 0$.

CONCLUDING REMARKS

After plotting curves such as those in Figs. 2 and 3 which describe the total appearance of the reliability degradation, we should be able to determine easily the design safe life and/or the inspection period so as to maintain the reliability level higher than the prescribed value of reliability. In the present study, the stress amplitude is fixed to a constant value, Z_0. However, it should be mentioned that, for those cases where the crack propagation resistance is kept constant and only the stress amplitude is a locally stationary random process, a similar and even simpler procedure can be applied.

Further, these curves give us insight into the relationship between inspection period and inspection guideline satisfying the requirement where the reliability level is to be of a value higher than the prescribed value, e.g. where the initial flaw takes on a given value.

In conclusion, the model proposed in the present study really has wide applicability for those areas where safety and reliability play an important role.

REFERENCES

[1] J. Nemec, J. Drexler and M. Klesnil, Development of fatigue cracks in real structures: Applications to aircraft design, *Proc. ICF-4*, **1**, 157–72 (1977).
[2] P. M. Besuner and A. S. Telelman, Probabilistic fracture mechanics, *Nucl. Eng. Design*, **43**-1, 99–114 (1977).
[3] D. G. Ford, *Reliability and Structural Fatigue in One-Crack Models*, ARL-Struc. report, No. 369, 1–18 (1978).
[4] J. L. Bogdanoff, A new cumulative damage model (Part 1), *ASME Trans., J. Appl. Mech.*, **45**-2, 241–5 (1978).
[5] J. L. Bogdanoff and W. Krieger, A new cumulative damage model (Part 2), *ASME Trans., J. Appl. Mech.*, **45**-2, 246–50 (1978).
[6] J. L. Bogdanoff, A new cumulative damage model (Part 3), *ASME Trans., J. Appl. Mech.*, **45**-4, 733–9 (1978).
[7] J. L. Bogdanoff and F. Kozin, A new cumulative damage model (Part 4), *ASME Trans., J. Appl. Mech.*, **47**-1, 40–4 (1980).

[8]   D. A. Virkler, B. M. Hillberry and P. K. Goel, The statistical nature of fatigue crack propagation, *ASME Trans., J. Eng. Mat. Technol.*, **101**-1, 148–53 (1979).

[9]   T. Sakai and T. Tanaka, Statistical study on distribution characteristics of fatigue crack propagation life of metallic materials, *Proc. 23rd Japan Congr. Mat. Res.*, **23**, 93–9 (1980).

[10] K. P. Oh, The prediction of fatigue life under random loading: a diffusion model, *Int. J. Fatigue*, **2**-3, 99–104 (1980).

[11] J. N. Yang, Statistical crack growth in durability and damage tolerance analyses, *Proc. AIAA/ASME/ASCE/AHS 22nd Struct., Struct. Dynam. and Mater. Conf.*, Atlanta, Georgia, 38 (1980).

[12] A. B. Lidiard, Applications of probabilistic fracture mechanics to light water reactor pressure vessels and piping, *Nucl. Eng. Design*, **60**, 49 (1980).

[13] J. Dufresne, Probabilistic application of fracture mechanics, *Proc. ICF-5*, No. 2, 517–31 (1981).

[14] R. Arone, A statistical model for fatigue fracture under constant-amplitude cyclic loading, *Eng. Fract. Mech.*, **14**, 189 (1981).

[15] F. Kozin and J. L. Bogdanoff, A critical analysis of some probabilistic models of fatigue crack growth, *Eng. Fract. Mech.*, **14**, 59 (1981).

[16] F. Bastenaire, H. P. Lieurade, L. Regnier and M. Truchon, Statistical analysis of fatigue crack growth, *ASTM STP 738*, 163–70 (1981).

[17] H. Ishikawa and M. Shinozuka, Probabilistic law of crack propagation, *Econ. Rev. Kagawa Univ.*, **54**-4, 1275–83 (1981).

[18] S. Tanaka, M. Ichikawa and S. Akita, Variability of m and C in fatigue crack propagation law $da/dN = C(\Delta K)^m$, *Int. J. Fract.*, **17**-5, R121–4 (1981).

[19] J. L. Bogdanoff and F. Kozin, On nonstationary cumulative damage models, *ASME Trans., J. Appl. Mech.*, **49**-1, 37–41 (1982).

[20] M. Ichikawa et al., Reliability-based study of fatigue life in consideration of crack initiation and propagation processes, *J. Soc. Mater. Sci., Japan*, **31**, 697 (1982).

[21] J. N. Yang, G. C. Salivar and C. G. Annis, Jr., Statistical modeling of fatigue-crack growth in a nickel-base superalloy, *Eng. Fract. Mech.*, **18**, 257 (1983).

[22] F. Kozin and J. L. Bogdanoff, On life behavior under spectrum loading, *Eng. Fract. Mech.*, **18**, 271 (1983).

[23] P. W. Hovey, J. P. Gallagher and A. P. Berens, Estimating the statistical properties of crack growth for small cracks, *Eng. Fract. Mech.*, **18**, 285 (1983).

[24] Y. K. Lin and J. N. Yang, On statistical moments of fatigue crack propagation, *Eng. Fract. Mech.*, **18**, 243 (1983).

[25] F. Kozin and J. L. Bogdanoff, On the probabilistic modeling of fatigue crack growth, *Eng. Fract. Mech.*, **18**, 623 (1983).

[26] K. Dolinski and G. I. Schueller, Methods of probabilistic fracture mechanics in the analysis of NPP components, *Proc. 8th Int. Conf. Struct. Mech. in Reactor Technol.*, Brussels, August 1985, Bergy.

[27] R. L. Stratonovich, *Topics in the Theory of Random Noise,* Vol. 1, Gordon and Breach, 89 (1963).
[28] H. Ishikawa and A. Tsurui, A stochastic model of fatigue crack growth in consideration of random propagation resistance, *Trans. Japan Soc. Mech. Engrs.,* **50**-454, 1309–15 (1984).
[29] A. Tsurui and H. Ishikawa, Theoretical study on the distribution of fatigue crack propagation life under stationary random loading, *Trans. Japan Soc. Mech. Engrs.,* **51**-461, 31–7 (1985).
[30] T. Nakagawa *et al.,* Reliability analysis of fatigue crack propagation life based on Markov chain model, *Proc. 4th Symp. Reliability Eng. in Design,* JSMS, 44–9 (1982).

Statistical *S–N* Testing Method with 14 Specimens: JSME Standard Method for Determination of *S–N* Curves

HAJIME NAKAZAWA

Department of Mechanical Engineering, Chiba University, 1-33 Yayoicho, Chiba 280, Tokyo

and SHOTARO KODAMA

Department of Mechanical Engineering, Tokyo Metropolitan University, 2-1-1 Fukazawa, Setagaya-ku, Tokyo 158, Japan

ABSTRACT

This paper introduces an important part of the standards established by the Japan Society of Mechanical Engineers (JSME): Standard Method of Statistical Fatigue Testing, JSME 002-1981. The *S–N* testing method, which uses 14 specimens, is a recommended procedure for determining the *S–N* curve based on statistical considerations with a relatively small number of specimens. This method assumes that the *S–N* curve consists of two linear parts, the slope part and the horizontal part, on either a semi-log or a log-log scale, and that the coefficient of variation in the slope part of the *S–N* curve is independent of the stress levels. Of the 14 specimens used in this method to determine the *S–N* curve, eight are used for the slope part and six for the horizontal part. The allotment of the number of specimens is based on obtaining nearly the same confidence interval for the estimation of the mean fatigue strength in the slope part and for the estimation of the fatigue limit by the staircase method in the horizontal part of the *S–N* curve. An example of the application of this method is given.

INTRODUCTION

It is well known that test data of fatigue strength of materials exhibit scattered results, even where the specimens are from the same lot and the applied cyclic stresses are equal. Therefore it is necessary to carry out statistical fatigue tests. Some countries have their own standards for fatigue tests more or less similar to the ISO Standard Method of Fatigue Testing. In Japan there are fatigue testing methods for standard

59

specimens, JIS (Japanese Industrial Standard) z 2273-2275, as well as for welded parts and plastics. Unfortunately, most of these have not been based on statistical methods because a statistical fatigue test requires a long time and a large number of specimens.

The Japan Society of Mechanical Engineers (JSME) planned to have a standard method of statistical fatigue test using a relatively small number of specimens, and decided to establish this as its own standard, since some difficulties were foreseen in establishing a national standard (JIS). A committee was formed in 1978* to investigate this problem, and as a consequence the Standard Method of Statistical Fatigue Testing, JSME S 002-1981 [1] (in Japanese), was established in 1981. The main purpose of this standard is to obtain statistical data (though admittedly minimal) under a certain level of confidence, with only a limited number of specimens. It is expected that by use of this standard the comparison of test data will be easier and that statistically significant fatigue data can be more easily accumulated.

This JSME standard includes S-N testing methods, fatigue life testing methods, testing methods of fatigue strength at finite fatigue lives, and fit tests. For the details see the Appendix. In this paper the S-N testing method with 14 specimens will be explained with an example.

DEFINITIONS

The S-N testing method. This is a method of determining an S-N curve using a total of only 14 specimens, i.e. six specimens for the fatigue limit and eight specimens for the finite fatigue life region, two each at four levels of stress amplitude.

Slope part of S-N curve. This is the part of the S-N curve which is determined by the data of finite fatigue life ($N = 5 \times 10^4$ to 1×10^7 cycles), not including the horizontal part of the S-N curve.

Horizontal part of S-N curve. This is the fatigue limit part of an S-N curve. In the case where $N_{cr} \geqslant 1 \times 10^7$ (N_{cr}, the critical number of stress cycles to failure, denotes the knee of the S-N curve), it is the part of the S-N curve where $N \geqslant N_{cr}$.

*Members of the committee: chairman, Hajime Nakazawa; co-chairmen, Satoshi Nishijima, Shotaro Kodama, Junichi Arai, Hiroshi Ishikawa, Masahiro Ichikawa, Takayoshi Ishiguro, Kazuo Uchino, Shinsuke Enomoto, Yuichi Kawada, Masanori Kawahara, Kōji Koibuchi, Toshiyuki Shimokawa, Hideo Kitagawa, Shinichi Tanaka, Takao Nakagawa, Isamu Yoshimoto, Masanori Kurita, Hiroshi Yano.

Non-fracture. In *S-N* testing, the experiments are to be stopped at $N = 10^7$ cycles; non-fracture indicates that the specimen is not fractured at this number of stress cycles.

PREPARATION OF TEST SPECIMENS

The preparation of test specimens is to be in accordance with Japanese Industrial Standards. In an actual experiment, it is advisable to prepare more than 14 specimens. Experimental results, as shown in Fig. 1, often require more than 14 specimens to complete the test due to inaccurate estimations of the initial stress level or of the stress step. In Fig. 1 the numerals beside the plots indicate the sequence of the test.

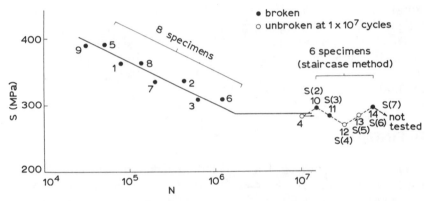

FIG. 1. A model of the *S-N* testing method with 14 specimens.

TEST PROCEDURE FOR SLOPE PART OF *S-N* CURVE

Estimate the fatigue strength at $N = 5 \times 10^4$ and $N = 1 \times 10^7$, namely S_A and S_B, from existing data for the same type of specimens under identical stress conditions with similar materials. Calculate the initial stress step from $d_I = (S_A - S_B)/3$ (use appropriate round number), followed by the initial stress level S_0 from $S_0 = S_A - d_I$ and the other stress levels $S_i = S_0 + id_I$ ($i = \pm 1, \pm 2, \ldots$).

Start the fatigue test at the initial stress level S_0 and continue the test, decreasing the stress level by the stress step d_I at stress levels $S_{-1}, S_{-2}, \ldots$, where $S_{-1} = S_0 - d_I$, $S_{-2} = S_{-1} - d_I$, using a specimen at each stress level until non-fracture is obtained. In the process of fatigue testing, revise the initial stress step d_1 and the stress levels S_i as necessary. Then, at the four stress levels above the one where non-fracture appeared, test other specimens so that two fractured data may be obtained at each of the four stress levels.

Plot the data on semi-log (S, log$N$) or log-log (log$S$, log$N$) paper to form the slope part of the S–N curve.

The slope part of the S–N curve (straight line for 50% survival is postulated in this test method) and the standard deviation of fatigue lives can be calculated by the following equations. Here, $\widehat{\ }$ denotes the estimated value and $^{-}$ denotes the arithmetic mean.

For the semi-log scale, the slope part of the S–N curve is given by

$$\log N = \widehat{\alpha}_1 + \widehat{\beta}_1 S$$

where

$$\widehat{\alpha}_1 = \overline{\log N} - \widehat{\beta}_1 \overline{S}$$

$$\widehat{\beta}_1 = \frac{\displaystyle\sum_{i=1}^{8} (S_i - \overline{S})(\log N_i - \overline{\log N})}{\displaystyle\sum_{i=1}^{8} (S_i - \overline{S})^2}$$

$$\overline{\log N} = \frac{1}{8} \sum_{i=1}^{8} \log N_i$$

$$\overline{S} = \frac{1}{8} \sum_{i=1}^{8} S_i$$

The estimated standard deviation of the logarithmic fatigue life $\sigma(\log N)$ is

$$\widehat{\sigma}(\log N) = \left[\frac{1}{6} \sum_{i=1}^{8} \{\log N_i - (\widehat{\alpha}_1 + \widehat{\beta}_1 S_i)\}^2 \right]^{1/2}$$

The estimated standard deviation of the fatigue strength $\sigma(S)$ is

$$\widehat{\sigma}(S) = \frac{1}{|\widehat{\beta}_1|} \widehat{\sigma}(\log N)$$

On the log-log scale, the slope part of the S-N curve is given by

$$\log N = \widehat{a}_2 + \widehat{\beta}_2 \log S$$

where

$$\widehat{a}_2 = \overline{\log N} - \widehat{\beta}_2 \overline{\log S}$$

$$\widehat{\beta}_2 = \frac{\displaystyle\sum_{i=1}^{8} (\log S_i - \overline{\log S})(\log N_i - \overline{\log N})}{\displaystyle\sum_{i=1}^{8} (\log S_i - \overline{\log S})^2}$$

$$\overline{\log N} = \frac{1}{8} \sum_{i=1}^{8} \log N_i, \qquad \overline{\log S} = \frac{1}{8} \sum_{i=1}^{8} \log S_i$$

The estimated standard deviation of the logarithmic fatigue life $\sigma(\log N)$ is

$$\widehat{\sigma}(\log N) = \left[\frac{1}{6} \sum_{i=1}^{8} \{\log N_i - (\widehat{a}_2 + \widehat{\beta}_2 \log S_i)\}^2 \right]^{1/2}$$

The estimated standard deviation of the fatigue strength $\sigma(\log S)$ is

$$\widehat{\sigma}(\log S) = \frac{1}{|\widehat{\beta}_2|} \widehat{\sigma}(\log N)$$

The estimated coefficient of variation of the fatigue strength in the slope part of the S-N curve is given by

$$\widehat{\eta}(S) = \left[\exp\left\{ 5 \cdot 302 \frac{\widehat{\sigma}^2(\log N)}{\widehat{\beta}_2^{\,2}} \right\} - 1 \right]^{1/2}$$

TEST PROCEDURE FOR HORIZONTAL PART OF S-N CURVE

The staircase method [2] is used with a small number of specimens. The first stress level $S(1)$ is the highest stress level where no specimen has been fractured in the test of the slope part of the S-N curve. However, if there are one or more stress levels where one specimen is fractured and another is not fractured, use the highest stress level among them as the first stress level $S(1)$. In any case, at least one non-fracture datum is already obtained; this stress level can be used as the first result of the staircase method without repeating the test.

The stress step for the staircase method, d_{II}, is determined from $d_{II} = \hat{\sigma}(S)$ or $d_{II} = S(1) \times \hat{\eta}(S)$. The stress levels after the first one are as follows:

For the second datum $S(2) = S(1) + d_{II}$

For the third to seventh data $S(j) = S(j - 1) \pm d_{II}$ $(j = 3, 4, 5, 6, 7)$

$(+)$ is chosen when the $(j - 1)$th specimen was not fractured, otherwise $(-)$ should be chosen. At the seventh stress level no testing is necessary, since the stress level is known, which suffices for the calculation of the fatigue limit.

The fatigue limit for 50% survival, S_w, or the fatigue strength at $N = 1 \times 10^7$ can be obtained by the following equation:

$$\hat{S}_w = \frac{1}{6} \sum_{j=2}^{7} S(j)$$

BALANCING THE NUMBER OF SPECIMENS IN THE SLOPE PART AND THE HORIZONTAL PART OF THE S-N CURVE [3]

In this S-N testing method (with 14 specimens) there are four stress levels where there should be two each with fractured data, i.e. eight data in total, and six data in the staircase method, making the total number of data 14.

The process behind the determination of this allocation was as follows. The committee came to the conclusion that the reasonable and practical number of specimens, n, for the S-N curve was approximately 15. The allotment of n to the slope part n_1 and to the horizontal part n_{II} is based on the consideration of making the resulting slope part and the horizontal part of the S-N curve well balanced in the statistical sense. The principle in this method is to obtain nearly the same confidence interval for the estimation of the fatigue strength in the slope part and for the estimation of the mean of the fatigue limit.

It is assumed that (a) the slope part of the S-N curve is linear on semi-log or full log scales, (b) the fatigue lives N are normally distributed on a log scale, and (c) $\sigma^2(\log N)$, the variance of $\log N$, is independent of the stress levels. The linear assumption on the semi-log scale is adopted here for explanation purposes. Therefore, the fatigue strength at N cycles, S, is normally distributed and $\sigma^2(S)$, the variance of S, is independent of the stress cycles, N. These assumptions are felt to be adequate as a first approximation from experimental results. When n_1 specimens are tested at stress levels $S_1, S_2, \ldots, S_{n_1}, n_1$ fatigue lives are obtained as $N_1, N_2, \ldots, N_{n_1}$, (it is permissible to have the same stress level among all S_i). Note that the suffixes here have nothing to do with those in the preceding part of this report.

The slope part of the $S-N$ curve can be obtained by the least-squares method, where $S_i = x_i$ and $\log N_i = y_i$. From this line, $\widehat{S}_N$, the estimated mean of the fatigue strength at y (= $\log N$), can be estimated by the following equation:

$$\widehat{S}_N = \bar{x} + \frac{1}{\widehat{\beta}}(y_i - \bar{y})$$

where

$$\widehat{\beta} = \frac{\Sigma(x_i - \bar{x})(y_i - \bar{y})}{\Sigma(x_i - \bar{x})^2}$$

and

$$\bar{x} = \frac{\Sigma x_i}{n_1}, \qquad \bar{y} = \frac{\Sigma y_i}{n_1}$$

The variance of $\widehat{S}_N$ can be calculated approximately by the following equation:

$$\sigma^2(\widehat{S}_N) \doteqdot \sum \left(\frac{\partial \widehat{S}_N}{\partial y_i}\right)^2_{y_i = \bar{y}_i} \sigma^2(y_i)$$

$$= \sigma^2(\log N) \sum \left(\frac{\partial \widehat{S}_N}{\partial y_i}\right)^2_{y_i = \bar{y}_i}$$

The variance of the fatigue strength at N cycles is

$$\sigma_1^2(S) = \frac{1}{\widehat{\beta}^2}\sigma^2(\log N)$$

From these equations

$$\sigma^2(\widehat{S}_N) \doteqdot \sigma_1^2(S)\left[\frac{1}{n_1} + \frac{(x - \bar{x})^2}{\Sigma(x_i - \bar{x})^2}\right]$$

This equation indicates that $\sigma^2(\widehat{S}_N)$ is larger when x is away from $\overline{x}$. The maximum value of $\sigma^2(\widehat{S}_N)$ in the range of stress levels is calculated as

$$\sigma^2(\widehat{S}_N)_{max} = k_I \frac{\sigma_I^2(S)}{n_I}$$

The fatigue limit, S_w, is also assumed to be normally distributed. When S_w is estimated by the staircase method, then $\sigma^2(S_w)$, the variance of the estimated fatigue limit S_w, is

$$\sigma^2(\widehat{S}_w) = k_{II} \frac{\sigma_{II}^2(S_w)}{n_{II}}$$

where $\sigma_{II}^2(S_w)$ is the variance of the fatigue limit.

To balance the accuracy of the estimated strength in the slope part and in the horizontal part of the S-N curve, it is proposed that the following relation be used:

$$\frac{\sigma(\widehat{S}_N)_{max}}{\sigma_I(S)} = \frac{\sigma(\widehat{S}_w)}{\sigma_{II}(S)}$$

If $\sigma_I(S) = \sigma_{II}(S)$, then the confidence interval of each part of the S-N curve has the same value at the intersection of each part. By substituting the values of $\sigma(\widehat{S}_N)_{max}$ and $\sigma(\widehat{S}_w)$, the number of specimens allotted for the slope part n_I and the horizontal part n_{II} can be expressed as

$$\frac{n_I}{n_{II}} = \frac{k_I}{k_{II}}$$

In the case of a test where each of m specimens is tested at l stress levels, which are separated by equal stress steps, k_I is calculated as

$$k_I = \frac{2(2l - 1)}{l + 1} = \frac{2(n_I - m)}{n_I + m}$$

Since Yoshimoto [2] has found that $k_{II} = 2$, the relation between n_I and n_{II} can be shown as follows:

$$n_{II} = \frac{l + 1}{2l - 1} n_I = \frac{n_I(n_I + m)}{2n_I - m}$$

This relation is shown in Fig. 2. As can be seen in Fig. 2, when $n = n_I + n_{II} \leqslant 15$ there can be two combinations.

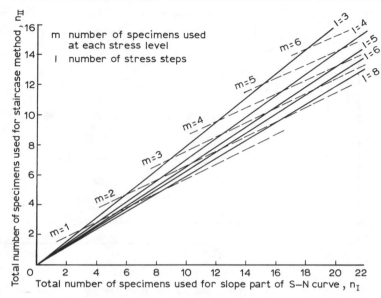

FIG. 2. Determination of numbers of specimens required.

(1) $l = 8, m = 1$, i.e. use one specimen each at eight stress levels for the slope part of the S-N curve, and five specimens for the staircase method.

(2) $l = 4, m = 2$, i.e. use two specimens each at four stress levels for the slope part, and six specimens for the staircase method.

However, the accuracy of the estimation of the slope part of the S-N curve is higher in the case of $l = 4, m = 2$, and $l = 4$ is still enough for the test of linearity of the data. Therefore, in this 14-specimen fatigue test method, $n_I = 8$, $n_{II} = 6$, and $n = 14$ was adopted as the standard. The calculation shown in Fig. 2 assumes that the stress levels are separated by equal stress steps, but it has also been confirmed by calculation that a slight difference in the stress steps does not affect the validity of $n_I = 8$ and $n_{II} = 6$.

A SAMPLE EXPERIMENT

Figure 3 shows experimental results comparing high-strength steel STK 55; material 1 has been embrittled and cracked by immersion into molten zinc, and material 2 has undergone the same heat history but

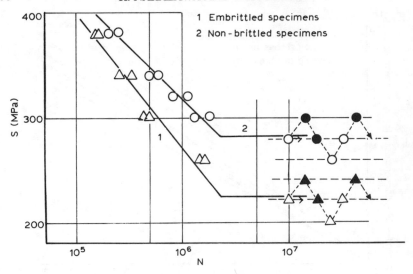

FIG. 3. An example of the application of the S–N testing method with 14 specimens; fatigue tests of a high-strength steel.

was not immersed in molten zinc and hence is not cracked. The cracks were confirmed easily by the red check method. The results were expressed as follows:

Material 1:

$$\log N = 8\cdot148 - 0\cdot007\,90S, \quad \hat{\sigma}(\log N) = 0\cdot1014,$$

$$\hat{\sigma}(S) = 12\cdot8\,\text{MPa}, \quad \hat{S}_w = 228\,\text{MPa}$$

Material 2:

$$\log N = 9\cdot346 - 0\cdot010\,54S, \quad \hat{\sigma}(\log N) = 0\cdot0681,$$

$$\hat{\sigma}(S) = 6\cdot5\,\text{MPa}, \quad \hat{S}_w = 283\,\text{MPa}$$

ACKNOWLEDGEMENT

The authors are grateful to the Japan Society of Mechanical Engineers for permission to translate a portion (sections 2 and 5.1) of the JSME Standard into English for this paper.

REFERENCES

[1] JSME, *Standard Method of Statistical Fatigue Testing*, JSME S 002 (1981).
[2] I. Yoshimoto, Rotating fatigue strength of rolled V-groove specimens, *Trans. Japan Soc. Mech. Engrs.,* **26**-167, 918 (1960).
[3] S. Tanaka, M. Ichikawa and T. Shimokawa, Statistical consideration on fatigue S-N testing, Papers JSME No. 817-2, 172 (1981).

APPENDIX

Contents of Standard Method of Statistical Fatigue Testing, JSME S 002-1981 (Japan Society of Mechanical Engineers).

Probabilistic Fracture Mechanics Investigation of Fatigue Crack Growth Rate

MASAHIRO ICHIKAWA

Department of Mechanical Engineering, University of Electro-Communications, 1-5-1 Chofugaoka, Chofu, Tokyo 182, Japan

ABSTRACT

Crack growth tests were conducted on 30 specimens of 2024-T3 Al alloy under identical conditions. Paris's law, $da/dN = C(\Delta K)^m$, was applied to each specimen, and the scatter of C and m among specimens was studied. To take into account the scatter of da/dN in the reliability analysis, (A) the inter-specimen variability method and (B) the intra-specimen variability method were examined. In (A), C and m are taken to be constant within a specimen, but to have specimen-to-specimen variability. In (B), C is taken to vary in space within a specimen and m not to vary in space. From the standpoint of (B), the specimen-to-specimen variability of C and m is simply a result of the spatial variation of C. In order to determine which method is better suited to the reliability analysis, the distribution of the crack propagation life was calculated based on each method and compared with the experimental distribution. It was also pointed out that crack growth tests under the condition of constant ΔK are useful to determine the better method. Furthermore, incorporation of the crack initiation life into the reliability analysis was considered.

INTRODUCTION

In structural reliability analyses based on probabilistic fracture mechanics, it is necessary to take into account the scatter of the crack growth rate, da/dN. However, an accepted method for this has not yet been established. The most straightforward approach seems to be to randomize C and/or m in Paris's law:

$$\frac{da}{dN} = C(\Delta K)^m \qquad (1)$$

There are two ways to accomplish this randomization:

(A) The inter-specimen variability method [1–5], in which C and/or

71

m are taken to be constant within a specimen but to have specimen-to-specimen variability.

(B) The intra-specimen variability method [5–8], in which C is taken to vary in space in a specimen and m not to vary in space. According to this second method, the specimen-to-specimen variability of C and m is simply a result of the spatial variation of C.

Equation (1) can be rewritten as

$$\frac{\mathrm{d}a}{\mathrm{d}N} = C_0\left(\frac{\Delta K}{K_0}\right)^m \tag{2}$$

where K_0 is a constant with the dimensions of the stress intensity factor. In terms of Eqn. (2), the inter-specimen variability method has the following variations:

Method (A1): m and C_0 are taken to be random variables independent of each other.

Method (A2): m is taken to be a random variable but C_0 is taken to be constant.

Method (A3): C_0 is taken to be a random variable but m is taken to be constant.

Through all three methods (A1), (A2) and (A3), K_0 is taken to be constant. From Eqns. (1) and (2), it follows that

$$C = C_0 K_0^{-m} \tag{3}$$

or

$$\log C = \log C_0 - m \log \Delta K \tag{4}$$

Hence, in terms of Eqn. (1), method (A1) assumes that both m and C are random variables and that there exists a certain degree of correlation between them. Similarly, method (A2) assumes that both m and C are random variables but that there is a linear relation between m and $\log C$, and method (A3) assumes that C is a random variable but m is constant.

The principal purpose of this paper is to review our efforts to find an appropriate probabilistic fracture mechanics method from among (A1), (A2), (A3), and (B), experimentally as well as theoretically. An approach different from fracture mechanics was proposed by Bogdanoff and Kozin [9], and followed by Nakagawa and co-workers [10].

EXPERIMENTAL INVESTIGATION [4]

The material used was 2024-T3 Al alloy plates of 1 mm thickness. Its chemical composition and mechanical properties are shown in Tables 1 and 2. The center-notched specimens were prepared by machining, as shown in Fig. 1. The longitudinal direction of the specimen coincides with the rolling direction.

TABLE 1
Chemical composition of the material tested (wt %)

Cu	Si	Fe	Mn	Mg	Cr	Zn	Zn + Ti	Ti
4·35	0·13	0·30	0·66	1·42	0·01	0·03	0·05	0·02

TABLE 2
Mechanical properties of the material tested

Proof stress (MPa)	Tensile strength (MPa)	Elongation (%)
327	478	19·9

Fatigue crack growth tests were carried out under axial loading using a closed-loop hydraulic testing machine. The stress range was $\Delta\sigma = 59\cdot4$ MPa, and the mean stress was $\sigma_m = 44\cdot8$ MPa. Hence the stress ratio was $R = 0\cdot2$. The repeating frequency was $f = 18\cdot4$ Hz. The temperature of the specimen was kept constant at 30°C by means of on/off control of a thermistor–relay–heating lamp system.

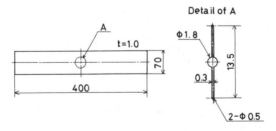

FIG. 1. Specimen geometry (mm).

Crack growth was monitored using a traveling microscope. The crack length was measured at the two tips of the crack on both sides of the specimen. The time interval of the measurement was approximately 5000 cycles. Thirty specimens were tested under identical conditions, and the results were analyzed statistically.

The crack length, a, is plotted against the number of cycles, N, in Fig. 2. $N = 0$ corresponds to the start of the test. The relation between the crack growth rate, da/dN, and the stress intensity range, ΔK, is shown in Fig. 3. da/dN was evaluated as $(a_{i+1} - a_i)/(N_{i+1} - N_i)$ where N_i and N_{i+1} are the number of cycles at which the ith and the $(i + 1)$th measurement of a were carried out, and a_i and a_{i+1} are the values of a at $N = N_i$ and $N = N_{i+1}$, respectively. da/dN thus calculated corresponds to the crack growth rate at the crack length of $(a_i + a_{i+1})/2$. ΔK was evaluated as

$$\Delta K = \Delta\sigma\sqrt{\pi a} \cdot \sec(\pi a/W) \tag{5}$$

where W is the specimen width. In Fig. 3 it should be noted that the low ΔK region does not represent the threshold behavior. The data points do not exist below $\Delta K = 9.5$ MPa $\cdot$ m$^{1/2}$ simply because the lowest applied ΔK in this experiment was 9.5 MPa $\cdot$ m$^{1/2}$. Paris's law, $da/dN = C(\Delta K)^m$, was applied to the data points of each specimen using

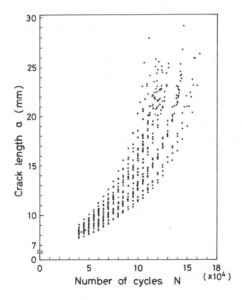

FIG. 2. Crack length, a, plotted against the number of repeated cycles, N.

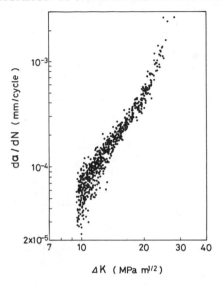

FIG. 3. Crack growth rate, da/dN, plotted against the stress intensity range, ΔK.

the method of least squares, and specimen-to-specimen variability of m and C was studied. Figures 4 and 5 show the distributions of m and $\log C$ as plotted on normal probability paper using the mean rank. It is seen that both m and $\log C$ show approximately normal distribution. Table 3 shows the mean value, the standard deviation, and the coefficient of

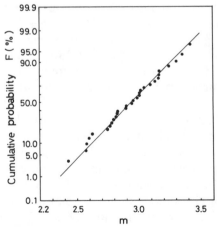

FIG. 4. Inter-specimen distribution of m (normal probability paper).

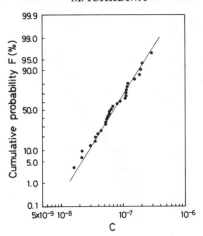

FIG. 5. Inter-specimen distribution of log C (log–normal probability paper).

TABLE 3
Statistics of m and $\log C$

	m	$\log C$
Distribution form	normal	normal
Mean value	2·939	−7·156
Standard deviation	0·247	0·310
Coefficient of variation	8·39 (%)	4·33 (%)

variation of both m and $\log C$. Figure 6 examines the correlation between m and $\log C$. It is seen that a strong negative correlation exists between m and $\log C$. Trends similar to those shown in Figs. 4–6 were observed by Sakai and Tanaka [3] and by Tanaka et al. [12]. The data in references [2] and [11] also show a similar trend, but it should be noted that these references are based on a mixture of data from various sources.

The results in Figs. 4–6 can be obtained if m and $\log C_0$ in Eqn. (2) are normally distributed. Applying Eqn. (4) to the data points in Fig. 6 and using the method of least squares, $K_0 = 17\cdot6$ MPa · m$^{1/2}$. Then, substituting this value of K_0 and the values of m and C into the equation $C_0 = CK_0^m$, the value of C_0 for each specimen was obtained. Figure 7 shows the distribution of $\log C_0$ as plotted on a normal probability scale. To the first approximation, the distribution of $\log C_0$ may be treated as normal.

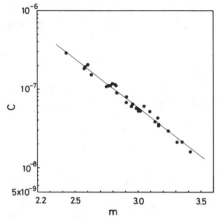

FIG. 6. Correlation between m and $\log C$.

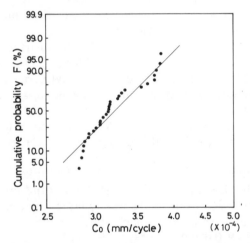

FIG. 7. Inter-specimen distribution of $\log C_0$ (log–normal probability paper).

COMPARISON OF INTER-SPECIMEN VARIABILITY METHOD WITH EXPERIMENT [5]

The experimental results described above suggest that method (A1) is suited to the data as far as the inter-specimen variability method is concerned. However, from the viewpoint of structural reliability analysis, it is necessary to examine whether the distribution of the crack propagation life can also be predicted accurately by method (A1). Methods (A2) and (A3) will also be examined for comparison purposes.

Calculation Procedure

Let N be the crack propagation life for the crack growth from $a = a_1$ to $a = a_2$. N is obtained by integrating Eqn. (2) as

$$N = \frac{K_0^m}{C_0} \int_{a_1}^{a_2} \frac{da}{(\Delta K)^m} \tag{6}$$

Substitution of Eqn. (5) into Eqn. (6) yields

$$N = \frac{1}{C_0} \left(\frac{K_0}{\Delta\sigma\sqrt{\pi}} \right)^m \int_{a_1}^{a_2} \left(a \sec \frac{\pi a}{W} \right)^{-m/2} da \tag{7}$$

The distribution of N was obtained by Monte Carlo simulation based on each of the three methods (A1)–(A3) as follows:

(1) Simulation based on method (A1): the distributions of m and $\log C_0$ were assumed as $m \sim N(2{\cdot}939, 0{\cdot}247)$ and $\log C_0 \sim N(-3{\cdot}49, 0{\cdot}0424)$ where the expression $x \sim N(\mu_x, \sigma_x)$ means that the variable x has a normal distribution with the mean, μ_x, and the standard deviation, σ_x.

(2) Simulation based on method (A2): m was taken to have the same normal distribution as given above. Log C_0 was taken to be a constant of value $-3{\cdot}49$.

(3) Simulation based on method (A3): m was taken to be a constant of value $2{\cdot}939$. Log C_0 was taken to be a random variable. The distribution of $\log C_0$ in this case was obtained by applying $da/dN = C_0(\Delta K/K_0)^m$ for each specimen. The resulting distribution is $\log C_0 \sim N(-3{\cdot}49, 0{\cdot}0502)$.

Other relevant input parameters in these simulations are: $K_0 = 17{\cdot}6\ \text{MPa} \cdot \text{m}^{1/2}$, $\Delta\sigma = 59{\cdot}4\ \text{MPa}$ and $W = 70\ \text{mm}$. a_1 was fixed at $9{\cdot}0\ \text{mm}$. a_2 was increased from $9{\cdot}5\ \text{mm}$ to $23{\cdot}0\ \text{mm}$ in increments of $0{\cdot}5\ \text{mm}$. 9999 replications of simulation were conducted for each case.

Results of Calculations

We first compared the distribution form obtained by Monte Carlo simulation with the experimental distribution form. In Fig. 8, the dots show the experimental data on log-normal probability paper. It is seen that the experimental distribution of N is log-normal. The lines show the distribution of N as obtained by Monte Carlo simulation based on method (A1). It is seen that method (A1) yields a log-normal distribution. It was also shown that methods (A2) and (A3) yield log-normal distributions.

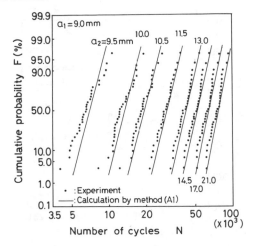

FIG. 8. Distribution of the crack propagation life, N, corresponding to crack growth from $a = a_1 = 9$ mm to $a = a_2$ (log-normal probability paper).

Next, let μ_N be the mean value of N. Figure 9 compares μ_N predicted by Monte Carlo simulation with the experimental value. It is seen that prediction by any of the three methods is in good agreement with the experimental results.

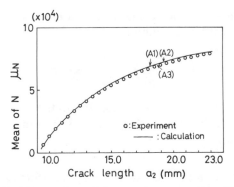

FIG. 9. Mean value of the crack propagation life, N.

Next, let η_N be the coefficient of variation of N. Figure 10 compares η_N as predicted by Monte Carlo simulation with the experimental value. It is seen that variation of η_N with a_2 as predicted by method (A1) is in reasonable agreement with the experimental results, but the other two methods (A2) and (A3) are not.

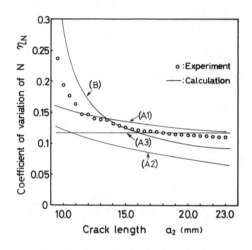

FIG. 10. Coefficient of variation of the crack propagation life, N.

The above comparisons show that method (A1) is promising as far as the inter-specimen variability method is concerned, with respect to the distribution form, to the mean value, and to the coefficient of variation of the crack propagation life, N.

Approximate Equations for μ_N and η_N Based on Method (A1)

Equation (7) can be rewritten as

$$N = \frac{W}{C_0}\left(\frac{K_0}{\Delta\sigma\sqrt{\pi W}}\right)^m \int_{\xi_1}^{\xi_2} [\xi \sec(\pi\xi)]^{-m/2}\,\mathrm{d}\xi \qquad (8)$$

where $\xi_1 = a_1/W$ and $\xi_2 = a_2/W$. According to the mean value theorem of integration, a value ξ_M ($\xi_1 < \xi_M < \xi_2$) exists that also satisfies the following equation:

$$\int_{\xi_1}^{\xi_2} [\xi \sec(\pi\xi)]^{-m/2}\,\mathrm{d}\xi = (\xi_2 - \xi_1)[\xi_M \sec(\pi\xi_M)]^{-m/2} \qquad (9)$$

From Eqns. (8) and (9),

$$N = \frac{a_2 - a_1}{C_0} \left[\frac{K_0}{\Delta\sigma\sqrt{\pi W}} \frac{1}{\sqrt{\xi_M \sec(\pi\xi_M)}} \right]^m \tag{10}$$

Hence

$$\log N = \log(a_2 - a_1) + m \log \left[\frac{K_0}{\Delta\sigma\sqrt{\pi W}} \frac{1}{\sqrt{\xi_M \sec(\pi\xi_M)}} \right] - \log C_0 \tag{11}$$

It was found that the function

$$G(m) \equiv \sqrt{\xi_M \sec(\pi\xi_M)} \equiv \left[\frac{1}{\xi_2 - \xi_1} \int_{\xi_1}^{\xi_2} \{\xi \sec(\pi\xi)\}^{-m/2} \, d\xi \right]^{-1/m} \tag{12}$$

is almost independent of m. Taking this fact into account, Eqn. (11) is rewritten as

$$\log N = \log(a_2 - a_1) + m \log \left[\frac{K_0}{\Delta\sigma\sqrt{\pi W}} \frac{1}{G(\mu_m)} \right] - \log C_0 \tag{13}$$

In the case of method (A1), the second and the third terms of the right-hand side of Eqn. (13) follow normal distribution. Hence, N given by Eqn. (13) is log-normally distributed. From Eqn. (13), the mean value and the variance of $\log N$ are obtained as

$$\mu_{\log N} = \log(a_2 - a_1) + \mu_m \log \left[\frac{K_0}{\Delta\sigma\sqrt{\pi W}} \frac{1}{G(\mu_m)} \right] - \mu_{\log C_0} \tag{14}$$

$$\sigma_{\log N}^2 = \sigma_m^2 \left[\log \left\{ \frac{K_0}{\Delta\sigma\sqrt{\pi W}} \frac{1}{G(\mu_m)} \right\} \right]^2 + \sigma_{\log C_0}^2 \tag{15}$$

Noting that N follows a log-normal distribution, the mean value of N is obtained as

$$\mu_N = \exp\left[2\cdot3\mu_{\log N} + \tfrac{1}{2}(2\cdot3\sigma_{\log N})^2\right] \tag{16}$$

An approximate equation for the coefficient of variation of N can be obtained by substituting Eqn. (15) into the following equation:

$$\eta_N = \sqrt{\exp(2\cdot3\sigma_{\log N})^2 - 1} \tag{17}$$

COMPARISON OF INTRA-SPECIMEN VARIABILITY METHOD
WITH EXPERIMENT [5]

It is possible that specimen-to-specimen variability of m and C comes from the spatial variation of C_0 in the specimen. Hence, it is natural to consider the intra-specimen variability method in which C_0 is taken to vary in space within a specimen and m not to vary in space.

Derivation of the distribution of the crack propagation life based on the intra-specimen variability method was studied by Kimura and Kunio [6], Ishikawa and Tsurui [7], and Lin and Yang [8]. In these investigations, correlation between the values of C_0 at different points in space was assumed to exist. Taking account of this correlation is important, as may be seen from the experimental results of da/dN by Virkler et al. [14].

In the next step, we derive the mean value and the coefficient of variation of the crack propagation life, N, in a simple way based on the intra-specimen variability method. We employ Kimura and Kunio's model [6] shown in Fig. 11. In this model, the crack propagation path is divided into the interval of the length, δ, and C_0 is taken to be constant within an interval but different in different intervals. The degree of correlation in spatial variation of C_0 is changed by changing δ. It should be noted that the interval within which C_0 is constant does not actually exist, but the essential feature of spatial variation of C_0 seems to be extracted by this model.

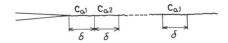

FIG. 11. Modeling of spatial variation of C_0 [6].

Let us divide the crack propagation path from $a = a_1$ to $a = a_2$ into n intervals. Let N_j be the number of cycles required for the crack to pass through the jth interval $a_1 + (j - 1)\delta \leqslant a \leqslant a_1 + j\delta$. N_j can be expressed as in Eqn. (7) as

$$N_j = \frac{1}{C_{0,j}} \left(\frac{K_0}{\Delta\sigma\sqrt{\pi}} \right)^m \int_{a_1 + (j-1)\delta}^{a_1 + j\delta} \left(a \sec \frac{\pi a}{W} \right)^{-m/2} da \qquad (18)$$

where $C_{0,j}$ is the value of C_0 in the jth interval. N is then expressed as

$$N = \sum_{j=1}^{n} N_j = \left(\frac{K_0}{\Delta\sigma\sqrt{\pi}}\right)^m \sum_{j=1}^{n} \frac{1}{C_{0,j}} \int_{a_1+(j-1)\delta}^{a_1+j\delta} \left(a \sec \frac{\pi a}{W}\right)^{-m/2} da \quad (19)$$

Using the first-order approximation, the mean value and the variance of N can be obtained as

$$\mu_N \simeq \left(\frac{K_0}{\Delta\sigma\sqrt{\pi}}\right)^m \frac{1}{\mu_{C_0}} \int_{a_1}^{a_2} \left(a \sec \frac{\pi a}{W}\right)^{-m/2} da \quad (20)$$

$$\sigma_N^2 \simeq \left(\frac{K_0}{\Delta\sigma\sqrt{\pi}}\right)^{2m} \frac{\sigma_{C_0}^2}{\mu_{C_0}^4} \delta \int_{a_1}^{a_2} \left(a \sec \frac{\pi a}{W}\right)^{-m} da \quad (21)$$

In the derivation of Eqn. (21), the following approximation was used:

$$\sum_{j=1}^{n} \left[\int_{a_1+(j-1)\delta}^{a_1+j\delta} \left(a \sec \frac{\pi a}{W}\right)^{-m/2} da\right]^2 \simeq \delta \int_{a_1}^{a_2} \left(a \sec \frac{\pi a}{W}\right)^{-m} da \quad (22)$$

From Eqns. (20) and (21), the coefficient of variation of N is obtained as follows:

$$\eta_N \simeq \eta_{C_0} \sqrt{\delta} \frac{\sqrt{\int_{a_1}^{a_2} \left(a \sec \frac{\pi a}{W}\right)^{-m} da}}{\int_{a_1}^{a_2} \left(a \sec \frac{\pi a}{W}\right)^{-m/2} da} \quad (23)$$

where μ_{C_0}, σ_{C_0} and η_{C_0} are the mean value, the standard deviation and the coefficient of variation of C_0, respectively. In the case of $W \gg a$ (or $\sec(\pi a/W) \simeq 1$), Eqn. (23) reduces to

$$\eta_N \simeq \eta_{C_0} \sqrt{\frac{\delta}{a_1}} \frac{m/2-1}{\sqrt{m-1}} \frac{\sqrt{1-\left(\dfrac{a_1}{a_2}\right)^{m-1}}}{1-\left(\dfrac{a_1}{a_2}\right)^{m/2-1}} \quad (24)$$

which is identical with Ishikawa and Tsurui's [7] expression derived by solving a Fokker–Planck equation.

Equations (20) and (21) hold good irrespective of the distribution form of C_0. In the case where C_0 follows a log-normal distribution, more exact equations can be obtained as follows:

$$\mu_N = \left(\frac{K_0}{\Delta\sigma\sqrt{\pi}}\right)^m \frac{1 + \eta_{C_0}^2}{\mu_{C_0}} \int_{a_1}^{a_2} \left(a \sec \frac{\pi a}{W}\right)^{-m/2} da \qquad (25)$$

$$\sigma_N^2 \simeq \left(\frac{K_0}{\Delta\sigma\sqrt{\pi}}\right)^{2m} \frac{(1 + \eta_{C_0}^2)^2 \sigma_{C_0}^2}{\mu_{C_0}^4} \delta \int_{a_1}^{a_2} \left(a \sec \frac{\pi a}{W}\right)^{-m} da \qquad (26)$$

η_N obtained from Eqns. (25) and (26) is the same as in Eqn. (23).

Let us compare μ_N given by Eqn. (20) and η_N given by Eqn. (23) with the experimental results. It should be noted that Eqn. (20) is identical with the expression of μ_N based on method (A3). Since μ_N obtained by method (A3) is in good agreement with experiment, as mentioned previously, it follows that Eqn. (20) also results in good agreement with experiment. Figure 10 compares Eqn. (23) with experiment. $\eta_{C_0}\sqrt{\delta/a_1}$ was taken to be $9\cdot43 \times 10^{-2}$. It is seen that the variation of η_N with a_2 predicted by the intra-specimen variability method is also in reasonable agreement with the experimental results.

A further analysis shows that the specimen-to-specimen variability of m and C resulting from the model in Fig. 11 is in good agreement with the experimental results of Figs. 4–6 [13].

INTER-SPECIMEN VARIABILITY METHOD VERSUS INTRA-SPECIMEN VARIABILITY METHOD

The above comparison shows that both the inter-specimen variability method and the intra-specimen variability method are in reasonable agreement with experiment as far as the distribution form, the mean value, and the coefficient of variation of the crack propagation life, N, are concerned. Hence it is necessary to examine the two methods from a different aspect. In this respect, it seems worthwhile to conduct crack growth tests under the condition of constant ΔK. By applying analysis of variance to results of such tests, the scatter of da/dN may be separated into two components: the scatter between specimens and the scatter within a specimen. If the former component is dominant, the inter-specimen variability method may be promising; if the latter component is dominant, the intra-specimen variability method may be promising. If these two components are comparable, it seems reasonable to consider a composite method in which the intra-specimen variability method and the inter-specimen variability method are combined. We are conducting crack growth tests under the condition of constant ΔK.

Furthermore, a theoretical analysis of the composite method has been undertaken [18].

DISTRIBUTION OF TOTAL FATIGUE LIFE [15–17]

When there are no initial sharp flaws, the fatigue life to final fracture, N_f, is composed of the crack initiation life, N_i, and the crack propagation life, N_p. Hence, the distribution of N_f is expressed as

$$F_f(N_f) = \int_0^{N_f} F_p(N_f - N_i) f_i(N_i) \, dN_i \tag{27}$$

if N_i and N_p are independent of each other, where $F_i(N_i)$, $F_p(N_p)$ and $F_f(N_f)$ are the distribution functions of N_i, N_p and N_f, respectively, and $f_i = dF_i/dN_i$. Figure 12 shows the experimental distributions of N_i, N_p and N_f [15]. This experiment was carried out on the same 2024-T3 Al alloy as was used in the crack growth test. Specimens of 52 × 11 × 1 mm plate with a 10 mm diameter circular center hole were subjected to reversed bending. The nominal bending stress range was 147 MPa. The

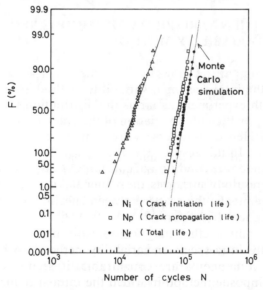

FIG. 12. Distributions of the crack initiation life, N_i, the crack propagation life, N_p, and the total life, N_f (log-normal probability paper).

crack initiation life, N_i, was defined as the number of cycles at which a crack of length 0·1 mm was observed. It is seen from Fig. 12 that both N_i and N_p approximately follow log-normal distributions. Using log-normal distributions for N_i and N_p, Monte Carlo simulation was conducted to calculate the distribution of $N_f = N_i + N_p$. As shown in Fig. 12, the calculated distribution of N_f is in good agreement with the experimental distribution of N_f.

In the above experiment, N_i is considerably smaller than N_p. If the applied stress is lowered, N_i will become comparable with, or larger than, N_p. Monte Carlo simulation for such cases yields Fig. 13. Hence, stress level dependence of the distribution of N_f is predicted, as shown in Fig. 14. Figure 14 is in agreement with the experimental results.

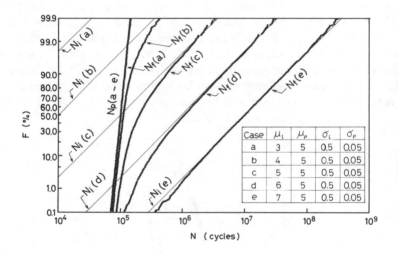

Case	μ_i	μ_p	σ_i	σ_p
a	3	5	0.5	0.05
b	4	5	0.5	0.05
c	5	5	0.5	0.05
d	6	5	0.5	0.05
e	7	5	0.5	0.05

FIG. 13. Result of Monte Carlo simulation for the distribution of the total fatigue life, N_f, when both the crack initiation life, N_i, and the crack propagation life, N_p, are log-normally distributed. (μ_i, σ_i) and (μ_p, σ_p) are (mean, standard deviation) of $\log N_i$ and $\log N_p$, respectively. The lines for N_p (a–e) are almost covered with the plots of N_f(a) (log-normal probability paper).

A similar stress-level dependence is expected in the stress corrosion cracking. Figure 15 shows the experimental distributions of the crack initiation life, t_i, the crack propagation life, t_p, and the total life, t_f, in stress corrosion cracking tests on notched specimens of Ni–Cr–Mo steel (JIS SNCM439) in 3·5% NaCl–water [16, 17]. In this case, both t_i and t_p

follow the two-parameter Weibull distributions. The distribution of $t_f = t_i + t_p$ as calculated by Monte Carlo simulation is in good agreement with the experimental distribution of t_f, as shown in Fig. 15.

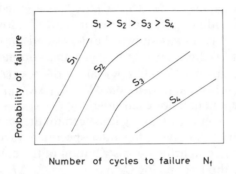

FIG. 14. Schematic illustration of the stress level dependence of the fatigue life distribution (log-normal or Weibull probability paper).

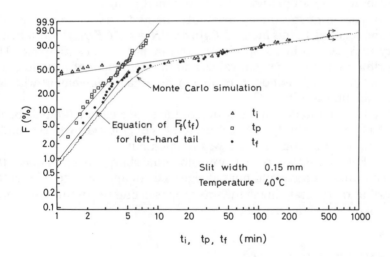

FIG. 15. Distributions of the crack initiation life, t_i, the crack propagation life, t_p, and the total life, t_f, in stress corrosion cracking of a Ni–Cr–Mo steel in 3·5% NaCl.

CONCLUSIONS

Statistical aspects of fatigue crack growth rate were investigated from the viewpoint of probabilistic fracture mechanics. The main conclusions obtained are as follows.

(1) When the equation $da/dN = C(\Delta K)^m$ is applied to the experimental data of individual specimens, m and C show specimen-to-specimen variability. It was shown that (i) m and $\log C$ are normally distributed, and (ii) there exists strong negative correlation between m and $\log C$. Rewriting $da/dN = C(\Delta K)^m$ as $da/dN = C_0(\Delta K/K_0)^m$, (i) and (ii) imply that m and $\log C_0$ in $da/dN = C_0(\Delta K/K_0)^m$ are normally distributed independently of each other.

(2) As the first method of taking into account the scatter of da/dN in structural reliability analysis, the inter-specimen variability method was examined by comparing the distribution of the crack propagation life, N. In this method, m and $\log C_0$ in $da/dN = C_0(\Delta K/K_0)^m$ were taken to be normally distributed independently of each other. The distribution form, the mean value, and the coefficient of variation of N predicted by this method were in reasonable agreement with the experimental results. The method in which only m or $\log C_0$ is taken to be normally distributed cannot predict the experimental results.

(3) The intra-specimen variability method was examined as the second method. In this method, C_0 in $da/dN = C_0(\Delta K/K_0)^m$ was taken to vary in space within a specimen, and m not to vary in space. The distribution form, the mean value, and the coefficient of variation of N predicted by this method were also in reasonable agreement with the experimental results.

(4) In order to determine which of the above two methods is better suited to the reliability analysis, usefulness of crack growth tests under the condition of constant ΔK was pointed out.

(5) For the case where there were no initial sharp flaws and hence the crack initiation life could not be neglected, an approach to predict the distribution of total fatigue life was proposed and the usefulness of this approach was demonstrated.

ACKNOWLEDGEMENTS

The author wishes to express his heartfelt thanks to Dr. S. Tanaka, President of the University of Electro-Communications, for his encouragement in this study. Thanks are also due to S. Akita and T. Takamatsu for useful discussions. The author is also grateful to M. Miyajima, M. Hamaguchi, T. Nakamura, T. Takura and J. Tsujimoto for their help. This study was partly supported by a grant-in-aid for scientific research from the Japanese Ministry of Education, Science and Culture.

REFERENCES

[1] H. Okamura, K. Watanabe and Y. Naito, *Reliability Approach in Structural Engineering*, A. M. Freudenthal *et al.*, Eds., Maruzen, 243 (1975).

[2] T. Kanazawa, H. Itagaki, S. Machida and Y. Kawamoto, *J. Soc. Naval Architects, Japan*, No. 146, 444 (1979) (in Japanese).

[3] T. Sakai and T. Tanaka, *J. Soc. Mater. Sci., Japan*, **28**, 92 (1979) (in Japanese).

[4] M. Ichikawa, M. Hamaguchi and T. Nakamura, *J. Soc. Mater. Sci., Japan*, **33**, 8 (1984) (in Japanese).

[5] M. Ichikawa and T. Nakamura, *J. Soc. Mater. Sci., Japan*, **34**, 378 (1985) (in Japanese).

[6] Y. Kimura and T. Kunio, *Trans. Japan Soc. Mech. Engrs.*, **42**, 1 (1976) (in Japanese).

[7] H. Ishikawa and A. Tsurui, *Trans. Japan Soc. Mech. Engrs.*, **A50**, 1309 (1984) (in Japanese).

[8] Y. K. Lin and J. N. Yang, *Eng. Fract. Mech.*, **18**, 243 (1983).

[9] J. L. Bogdanoff and F. Kozin, *J. Appl. Mech., Trans. ASME*, **47**, 40 (1980).

[10] Y. Shimada, T. Nakagawa and H. Tokuno, *J. Soc. Mater. Sci., Japan*, **33**, 475 (1984) (in Japanese).

[11] W. G. Clark, Jr. and S. J. Hudak, Jr., *J. Test. Eval.*, **3**, 454 (1975).

[12] S. Tanaka, M. Ichikawa and S. Akita, *Int. J. Fracture*, **17**, R121 (1981).

[13] M. Ichikawa and S. Akita, *J. Soc. Mater. Sci., Japan*, **35**, 1177 (1986) (in Japanese).

[14] D. A. Virkler, B. M. Hilberry and P. K. Goel, *J. Eng. Mat. Tech., Trans. ASME*, **101**, 148 (1979).

[15] M. Ichikawa, T. Takura and S. Tanaka, *J. Soc. Mater. Sci., Japan*, **31**, 697 (1982) (in Japanese).

[16] M. Ichikawa, T. Takura and S. Tanaka, *Bull. Japan Soc. Mech. Engrs.*, **26**, 1857 (1983).

[17] M. Ichikawa, T. Takura and J. Tsujimoto, *Bull. Japan Soc. Mech. Engrs.*, **26**, 2039 (1983).

[18] M. Ichikawa and S. Akita, *J. Soc. Mater. Sci., Japan*, **35**, 1272 (1986) (in Japanese).

Fatigue Life Prediction and Material Evaluation Based on a Statistics-of-Extremes Analysis of the Maximum Length Distributions of Multiple Small Cracks

YUUJI NAKASONE

National Research Institute for Metals, Tsukuba Labs., 1-2-1 Sengen, Sakura-mura, Niihari-gun, Ibaraki 305, Japan

TSUYOSHI SHIMAZAKI

Graduate School, Yokohama National University, 153 Tokiwadai, Hodogaya-ku, Yokohama, Kanagawa 240, Japan

MINEAKI IIDA

Toshiba Electric Co. Ltd, 8 Shinsugita-machi, Isogo-ku, Yokohama, Kanagawa 235, Japan

and HIDEO KITAGAWA

Department of Engineering, Yokohama National University, 153 Tokiwadai, Hodogaya-ku, Yokohama, Kanagawa 240, Japan

ABSTRACT

The present paper investigates fatigue life prediction and material evaluation by a statistics-of-extremes analysis of the maximum length distributions of multiple small cracks in steel specimens under rotating bending fatigue in air. Three different types of steel having the same strength level were investigated: a high carbon steel (JIS S45C), a non-degraded steel, and a sigma-phase-degraded steel. A maximum length distribution was determined by the surface length of the largest crack, $2a_{max}$, in each of 50 narrow band regions. The narrow band regions were 0·2 mm wide in the longitudinal direction of the specimen by 10 mm long in its circumferential direction. These dimensions were chosen since, even when cracks coalesced, they grew in the circumferential direction of the specimen lying within narrow band regions of less than 0·2 mm width. The $2a_{max}$ distributions were regarded as three-parametric Weibull distributions in all three steels tested. The Weibull parameters and return periods, T, were calculated from the $2a_{max}$ distributions, and their variations with cycle ratio, N/N_f, or number of load cycles, N, were obtained. The log-log T vs. N/N_f or N relationship was found to be given by a straight line for a given critical surface length, $2a_c$, in all the steels tested, and the application of the log-log T vs. N/N_f or N relationship to fatigue life prediction was proposed. The results showed that the relationship predicted the fatigue lives of the steels with reasonable accuracy.

INTRODUCTION

Problems caused by multiple small cracks have recently aroused increasing interest in obtaining a deeper understanding of the mechanisms of corrosion fatigue and low-cycle fatigue of structural materials and in establishing an effective method to predict their fatigue lives [1–15].

Multiple small cracks are only on the order of $0 \cdot 01$–$0 \cdot 5$ mm in surface length, and are often initiated in a densely distributed array over material surfaces; nevertheless, they grow and coalesce with one another to form macrocracks large enough to cause fracture [6, 13, 14].

Several statistical investigations have been made on the problems of multiple small cracks [1–7, 12–14], and some fatigue life prediction methods have been proposed [4, 6, 13]; for example, a Monte Carlo simulation analysis has been applied to life prediction for a fatigue process in which multiple small cracks play a significant role [6, 13]. The method has attempted to predict fatigue lives of unnotched materials by computer simulation of the initiation, growth, and coalescence behavior of multiple cracks. This method, however, requires a vast amount of data on multiple cracks and is overly time-consuming.

Multiple cracks clearly have a random nature in terms of initiation site, size, and growth rate; nevertheless, these cracks tend to grow in the same direction, i.e. the direction perpendicular to the principal stress axis; even when the cracks coalesce, the fracture line is often formed within a narrow band region on a specimen surface.

Taking these points into consideration, the present paper proposes a fatigue life prediction method based on a statistics-of-extremes analysis of maximum surface length distributions of multiple small cracks, which are determined by the surface length of the largest crack in each of 50 narrow band regions in a given portion of a specimen surface. The present method can greatly reduce laborious work of measurements; e.g. the above-mentioned Monte Carlo simulation analysis requires measurements of surface lengths and initiation sites of 400 to over 1000 small cracks per cm^2, whereas the proposed method requires measurements of surface lengths of only 50 cracks per cm^2. In this paper, fatigue life predictions will be made for three different types of steel having the same strength level.

EXPERIMENTAL PROCEDURES

Materials and Specimens Tested

Materials tested were a high carbon structural steel (JIS S45C) normalized at 850 °C and two types of stainless steel similar to ASTM CF 8M having the same tensile strength level as the S45C, i.e. a stainless steel containing approximately 30 wt% ferrite, and that in which a part of the 30 wt% ferrite was transformed into approximately 10 wt% sigma phase. The latter simulates a stainless steel degraded by sigma phase precipitation, so that in this paper the latter stainless steel is called a degraded stainless steel, whereas the former is a non-degraded stainless steel.

The chemical composition and the mechanical properties of the materials tested are listed in Tables 1 and 2. The specimens used in this study were of unnotched round bar types as shown in Fig. 1. The surface of the parallel section of a specimen was carefully polished with emery paper prior to a test.

TABLE 1
Chemical composition of the steels tested (wt%)

Material	S	Si	Mn	P	S	Cu	Ni	Cr	Mo
High carbon steel (S45C)	0·45	0·23	0·68	0·02	0·02	0·12	0·05	0·09	—
Stainless steels	0·05	1·1	0·86	0·02	0·01	—	10	22	2·4

TABLE 2
Mechanical properties of the steels tested

Material	Yield[a] strength (MPa)	Tensile strength (MPa)	Elongation (%)	Reduction of area (%)
High carbon steel (S45C)	405	648	30	51
Non-degraded stainless steel	254	628	60	64
Sigma-phase degraded stainless steel	250	637	54	48

[a]Lower yield point for S45C steel and 0·2% proof stress for stainless steels.

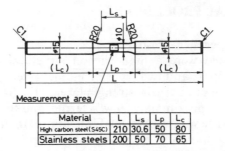

Material	L	L_s	L_p	L_c
High carbon steel(S45C)	210	30.6	50	80
Stainless steels	200	50	70	65

FIG. 1. Specimen geometries (all dimensions in mm).

Fatigue Test Procedures

Fatigue tests were conducted on an Ono-type rotating bending fatigue tester at room temperature in air. The cyclic loading rate was 2850 rpm.

The surface replication technique [14] was used for measuring the density, the surface lengths and, if necessary, the locations of multiple small cracks initiated on the smooth surface of a specimen at various stages of the fatigue process; i.e. a fatigue test was repeatedly paused at adequate number-of-load-cycles intervals in order to take replicas of multiple cracks on a specimen until it was finally fractured. The surface lengths and the locations of the cracks on the replica films so obtained were measured with an ocular micrometer attached to an optical microscope. The micrographs of the replica films were also used for the measurements.

Multiple small surface cracks measured in this study were those within a square region 10 mm × 10 mm on a smooth specimen surface located at the center of the specimen span (see Fig. 1). This measurement area was carefully set at the same site on the surface for each specimen, although it was copied on different replica films at different numbers of load cycles.

The distribution of the maximum surface lengths of the cracks was also determined by measuring the surface length of the largest crack in each of 50 sub-areas 0·2 mm × 10 mm; this is shown in Fig. 2. The size of the sub-areas was determined based on the observation that cracks grew in the circumferential direction of a specimen, and that, even when the cracks coalesced, they lay within narrow band regions having a width smaller than 0·2 mm.

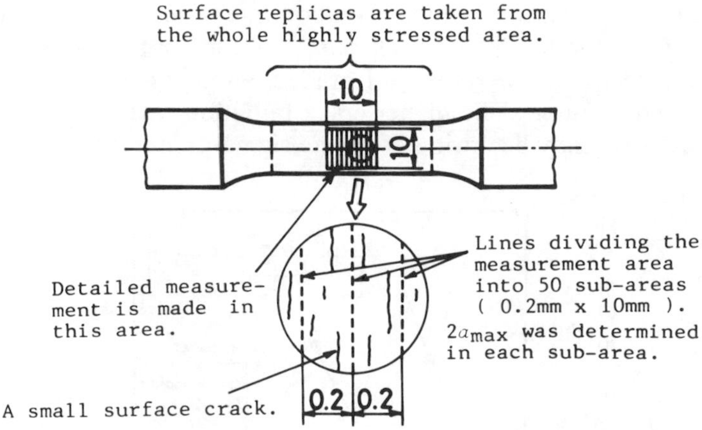

FIG. 2. Illustration of the sampling of data on crack lengths by the use of surface replica films (all dimensions in mm).

RESULTS AND DISCUSSION

Variation of the Number of Multiple Small Surface Cracks with the Number of Load Cycles

Figure 3 shows the variations in the number of multiple small surface cracks initiated in the measurement area 10 mm × 10 mm with the number of load cycles, N, in the three types of steel tested. In the S45C steel, the number of cracks per cm^2 increased remarkably for $N/N_f > 0.8$ where N_f is the number of load cycles to fracture, eventually reaching as high a value as approximately $400/cm^2$ in the last stage of the fatigue process. For the stainless steels, on the other hand, the crack density per cm^2 was more than double that in the S45C steel; i.e. it reached over $1000/cm^2$ in the last stage of the fatigue process in both the non-degraded and the degraded stainless steels. The difference in crack density may be explained by the difference in the severity of the deformation between the S45C steel and the stainless steels; i.e. in the stainless steels, since the applied stress range σ_a was about 1·4 times higher than the 0·2% proof stresses, plastic deformation occurred and may have accelerated crack initiation to bring about higher crack density than in the S45C steel, in which σ_a was lower than its lower yield point and so the plastic deformation was considered not so severe as in the stainless steels.

Figure 3 also shows that the initiation rate of multiple small cracks was highest for the S45C steel, next highest for the non-degraded stainless steel, and lowest for the degraded stainless steel at nearly equal applied stress range. The higher crack initiation rate in the degraded stainless steel may be attributed to the sigma phase.

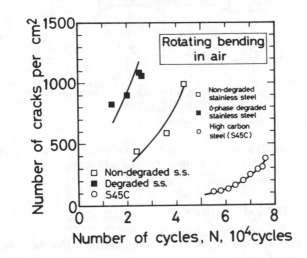

FIG. 3. Variations of the number of multiple small cracks initiated at the surface of the steels tested.

Statistical Distribution of Surface Lengths and Maximum Lengths of Multiple Small Cracks

Figures 4–7 show the statistical distribution of the surface lengths, $2a$, of all the cracks observed in the measurement area 10 mm × 10 mm, and that of the surface lengths of the 50 largest cracks in the 50 sub-areas 0·2 mm × 10 mm, $2a_{max}$, in the S45C steel and the stainless steels, respectively. The statistical test of goodness of fit showed that both $2a$ and $2a_{max}$ in the three steels tested could be regarded as following three-parametric Weibull distributions.

The curves in Figs. 4–7 shift rightwards as the load cycle ratio, N/N_f, increases; such a tendency implies the growth of multiple small cracks. The variations of the curves with N/N_f are larger for $2a_{max}$ than for $2a$ in the materials tested in this study, so the statistical distributions of $2a_{max}$, which can be obtained much more easily than those of $2a$, can better describe the progress of the fatigue fracture processes in the steels.

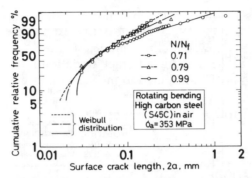

FIG. 4. Distributions of surface crack lengths, $2a$, in rotating bending fatigue of a high carbon steel (S45C) in air.

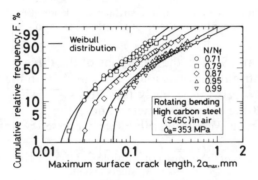

FIG. 5. Distributions of maximum surface crack lengths, $2a_{max}$, in rotating bending fatigue of a high carbon steel (S45C) in air.

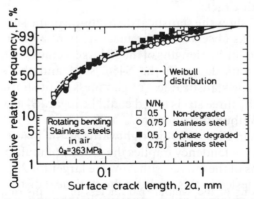

FIG. 6. Distributions of surface crack lengths, $2a$, in rotating bending fatigue of stainless steels in air.

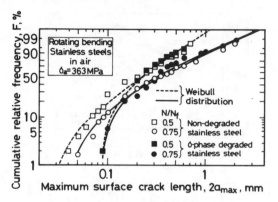

FIG. 7. Distributions of maximum surface crack lengths, $2a_{max}$, in rotating bending fatigue of stainless steels in air.

Figures 8 and 9 show the variations of the three parameters of the statistical distributions of $2a$ and $2a_{max}$ with N/N_f in the S45C steel and the stainless steels, respectively (see Figs. 4–7), where α is the scale parameter, β the shape parameter, and γ the location parameter of three-parametric Weibull distributions expressed as follows:

$$F(x) = 1 - \exp\{-[(x - \gamma)/\alpha]^{\beta}\} \tag{1}$$

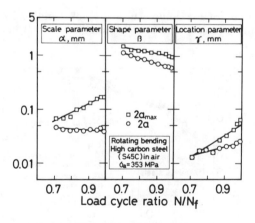

FIG. 8. Variations of the Weibull parameters in the distributions of surface crack lengths, $2a$, and maximum surface crack lengths, $2a_{max}$, with load cycle ratio, N/N_f, in a high carbon steel (S45C).

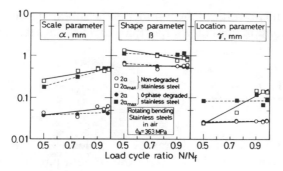

FIG. 9. Variations of the Weibull parameters in the distributions of surface crack lengths, $2a$, and maximum surface crack lengths, $2a_{max}$, with load cycle ratio, N/N_f, in a non-degraded and a degraded stainless steel.

These figures show that the logarithm of each of the three parameters varied linearly with N/N_f or remained constant, and that the variations of the parameters are larger for $2a_{max}$ than for $2a$.

Fatigue Life Prediction by a Statistics-of-Extremes Analysis of the Maximum Surface Crack Length Distributions

This section describes the fatigue life prediction of the three steels tested by the use of the return-period concept, which has been applied to the prediction of the time to leak in pipelines, in oil storage tanks, etc. caused by corrosion pits [16-19]. The present study attempts to apply this concept to the fatigue life prediction of steels fractured by multiple small cracks.

The return period, T, in the present case, gives the average size of an area, in which a crack having a surface length larger than a given critical value, $2a_c$, is expected to appear, normalized by the measurement sub-area 0.2 mm × 10 mm. The value of T can be calculated from the distribution function of $2a_{max}$, $F(2a_{max})$, for a given critical surface length, $2a_c$, by the following equation:

$$T = 1/[1 - F(2a_c)] \tag{2}$$

since the probability, p, that one crack having a surface length larger than $2a_c$ is to appear in one of n measurement sub-areas is

$$p = [1 - F(2a_c)][F(2a_c)]^{n-1} \equiv (1 - F)F^{n-1} \tag{3}$$

and thus the mean value of the number of the sub-areas, one of which is expected to contain one crack larger than $2a_c$ in surface length, becomes [20]

$$\sum_{n=1}^{\infty} np = \sum_{n=1}^{\infty} n(1-F)F^{n-1} = 1/(1-F) \equiv T \qquad (4)$$

When the value of T decreases to a level T^*, equal to that corresponding to the surface area of the smooth parallel section of a specimen, a crack larger than $2a_c$ can be expected to appear in the specimen. Therefore, assuming that fracture should occur when a crack larger than $2a_c$ appears in a specimen, the number of load cycles at which the value of T is equal to T^*, the ratio of the area of the parallel section of a specimen to that of the measurement sub-area 2 mm², can be regarded as the predicted fatigue life, N_f^*, of the specimen.

From Fig. 10, assuming the values of $2a_c$ in the present tests to be 1·4 mm for the S45C steel and 7 mm for the stainless steels, the variations of T with N/N_f were obtained as shown in Fig. 11 for the three steels. Figure 11 shows that the values of log-log T decreased linearly with N/N_f in the steels for the $2a_c$ values above, implying that the degree of the danger of fracture increases drastically with N/N_f or N.

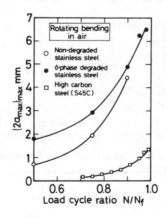

FIG. 10. Variations of the largest surface crack length of 50 maximum surface crack lengths in the measurement area 10 mm × 10 mm with load cycle ratio, N/N_f.

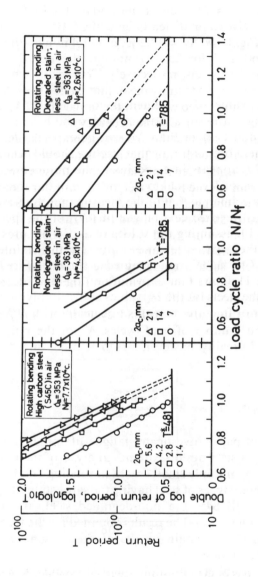

FIG. 11. Fatigue life prediction by the log-log T vs. N/N_f diagrams of the three steels tested; T is the return period, T^* is the value of T corresponding to the total area of the smooth parallel portion of a specimen, and $2a_c$ is the critical surface crack length.

The values of T^* in the present case were approximately 481 for the S45C steel and 785 for the stainless steels. The N/N_f value at an intersection of a $T = T^* = $ const. line and the log-log T vs. N/N_f line or its extrapolation gives a predicted fatigue life, N_f^*/N_f or N_f^*, for a given $2a_c$ value. From Fig. 11, $N_f^*/N_f \approx 1.0$ for $2a_c = 1.4$ mm for the S45C steel and $N_f^* \approx 0.88$ and 1.15 for $2a_c = 7$ mm for the non-degraded and the degraded stainless steels, respectively. The results whereby the N_f^*/N_f values are nearly equal to unity mean that fatigue lives, N_f^*, predicted by the present method agree well with the fatigue lives, N_f, obtained from the present tests with reasonable accuracy.

Figure 11 also shows that the present method of fatigue life prediction was not greatly affected by the $2a_c$ value, since even when the $2a_c$ value became, for example, three times larger than those estimated by Fig. 10, i.e. $2a_c = 4.2$ mm for the S45C steel and 21 mm for the stainless steels, the N_f^*/N_f values changed little, i.e. $N_f \approx 1.13$ for the S45C steel, 1.0 for the non-degraded stainless steel, and 1.4 for the degraded stainless steel. The difference between N_f^* and N_f values for the degraded stainless steel, however, was somewhat larger compared with that obtained for the other two steels; such a tendency may imply that the sigma phase affected the right foot of the distribution functions of $2a_{max}$ to produce cracks too large to be predicted.

Throughout this paper, the load cycle ratio, N/N_f, has been used instead of the number of load cycles, N, for the purpose of ease of comparison. In reality, however, N is to be used, since the real N_f is not known.

CONCLUSION

The present paper investigates fatigue life prediction and material evaluation by a statistics-of-extremes analysis of the maximum length distributions of multiple small cracks in rotating bending fatigue in air for three different types of steel having the same strength level: i.e. a high carbon steel (JIS S45C), a non-degraded steel, and a sigma-phase degraded stainless steel. The results obtained for the three types of steel were compared. The main results of the present investigation are summarized as follows.

(1) A large number of multiple small surface cracks were observed to be initiated in all the steels tested, i.e. 400 cracks per cm^2 in the S45C steel and over 1000 cracks per cm^2 in the stainless steels. The large number of

cracks enabled fatigue life prediction based on the statistics of extremes.

(2) Multiple small cracks measured were on the order of 0.01–0.5 mm in surface length, and were densely distributed over the surfaces of the steels tested; nevertheless, they grew and coalesced with one another to form macrocracks large enough to cause fracture of the steels.

(3) The distributions of the maximum crack lengths in 50 narrow-band regions on specimen surfaces 0.2 mm $\times$ 10 mm, with larger sides parallel to the crack growth direction, and the critical length, $2a_c$, were determined based on observation. Not only the statistical distribution of the surface lengths, $2a$, of all the cracks observed in the measurement area 10 mm $\times$ 10 mm, but that of the surface lengths of the 50 largest cracks in the 50 narrow-band regions, $2a_{max}$, were regarded as following three-parametric Weibull distributions.

(4) The logarithms of the Weibull parameters varied linearly with the load cycle ratio, N/N_f (or the number of load cycles, N), or remained constant. The variations of the $2a_{max}$ distributions were larger than those of the $2a$ distributions. Thus the $2a_{max}$ distributions, which can be measured much more easily than the $2a$ distributions, can better describe the progress of the fatigue fracture processes in the steels tested.

(5) The relationship between the double logarithm of the return period, log-log T, and N/N_f or N was found to be given by a straight line for a given critical surface length, $2a_c$, in all the steels tested, and the application of this relationship to fatigue life prediction was proposed. The results showed that the relationship predicted the fatigue lives of the tested steels with reasonable accuracy.

REFERENCES

[1] S. Ishihara, K. Shiozawa and K. Miyao, Preprint of Japan Soc. Mech. Engrs., No. 780-13, 13 (1978) (in Japanese).

[2] S. Sasaki and Y. Koshiji, *Proc. 16th Symp. Fatigue*, Japan Soc. Mech. Engrs., No. 790-9, 13 (1979) (in Japanese).

[3] K. Tokaji, Z. Ando, M. Sugimoto and N. Nakano, *J. Soc. Mater. Sci., Japan*, **30**-328, 15 (1981) (in Japanese).

[4] T. Tanaka, T. Sasaki and K. Okada, *J. Soc. Mater. Sci., Japan*, **30**-332, 483 (1981) (in Japanese).

[5] S. Ishihara, K. Shiozawa and K. Miyao, Preprint of Japan Soc. Mech. Engrs., No. 810-11, 161 (1981) (in Japanese).

104   Y. NAKASONE, T. SHIMAZAKI, M. IIDA AND H. KITAGAWA

[6]   H. Kitagawa, T. Fujita and K. Miyazawa, *ASTM STP 642*, 98 (1978).
[7]   H. Kitagawa, K. Tsuji, Y. Nakasone and T. Fujita, Preprint of Japan Soc.
       Mech. Engrs., No. 780-4, 104 (1978) (in Japanese).
[8]   H. Kitagawa, S. Takahashi, C. M. Suh and S. Miyashita, *ASTM STP 675*,
       420 (1978).
[9]   H. Kitagawa and S. Takahashi, *Trans. Japan Soc. Mech. Engrs.*, **45**-399, 1289
       (1979) (in Japanese).
[10]   M. H. El Haddad, K. N. Smith and T. H. Topper, *Trans. ASME, Ser. H,* **101**,
       42 (1979).
[11]   S. Usami and H. Kimoto, Preprint of Japan Soc. Mech. Engrs., No. 810-2,
       200 (1981) (in Japanese).
[12]   H. Kitagawa and Y. Nakasone, Preprint of Japan Soc. Mech. Engrs., No.
       817-1, 22 (1981) (in Japanese).
[13]   H. Kitagawa and Y. Nakasone, *J. Soc. Mater. Sci., Japan*, **33**-364, 14 (1984)
       (in Japanese).
[14]   H. Kitagawa, Y. Nakasone and M. Shimodaira, *Trans. Japan Soc. Mech.
       Engrs.,* **51**-463, 587 (1985) (in Japanese).
[15]   H. Kitagawa, Y. Nakasone and M. Shimodaira, *Trans. Japan Soc. Mech.
       Engrs.,* **51**-464, 1026 (1985) (in Japanese).
[16]   P. M. Aziz, *Corrosion,* **12**, 495 (1956).
[17]   N. Masuko, *Boushoku Gijutsu (Corr. Eng.)*, **21**, 347 (1972) (in Japanese).
[18]   T. Shibata, *Boushoku Gijutsu (Corr. Eng.)*, **27**, 23 (1978) (in Japanese).
[19]   Y. Ishikawa, *Boushoku Gijutsu (Corr. Eng.)*, **28**, 278 (1979) (in Japanese).
[20]   Z. Tanabe, *Sumitomo Light Met. Reps.,* **25**, 55 (1984) (in Japanese).

Statistical Studies on Fracture Toughness of Steels

NAMIO URABE

Technical Research Center, Nippon Kokan KK, 1-1 Minamiwatarida-cho, Kawasaki-ku, Kawasaki 210, Japan

ABSTRACT

The distribution of the critical crack tip opening displacement, δ_c, of a low alloy steel for welded structures (JIS G3106 SM50B) was experimentally obtained at $-70\,°C$. Based on the fractographic observation and the stress analysis made by McMeeking, the probabilistic nature of the fracture toughness of this steel was theoretically analyzed. The analysis showed that δ_c obeyed the Weibull distribution and that the shape parameter was 2. This prediction coincided well with the experimental results. A method was thus proposed to estimate the large-specimen toughness from the results of small-specimen fracture toughness tests as $\overline{\delta}_{cL} = \overline{\delta}_{cs}(T_s/T_L)^{1/2}$, where $\overline{\delta}_{cL}$ and $\overline{\delta}_{cs}$ are the average toughness and T_L and T_s the thickness of the large and the small specimen, respectively. The beta distribution was proposed and examined to fit the toughness distribution, especially for the lower toughness region, so that a more accurate failure probability would be obtained.

INTRODUCTION

Flaws in material and welded joints occasionally affect the safety and reliability of structures and consequently reduce the life of a structure. Analyses on accidental failures of several pressure vessels [1], offshore structures and petrochemical process reactors [2] have shown that the crack sizes which caused the failures were sufficiently large to be detected by non-destructive test techniques. Moreover, the fracture toughness values of the cracks calculated according to the flaw acceptance standards [3-6] were larger than the critical fracture toughness of the material. The cracks which caused the accidental failures were unfortunately overlooked by the non-destructive investigation. If the cracks had been detected and their effects on the structural safety had been correctly estimated, the accidents might have been

105

avoided. However, none of the non-destructive test techniques is perfect, since the result can only be as accurate as the available tools, including the degree of experience of the operators involved. The critical fracture toughness of a material is usually determined by conducting fracture toughness tests on three test pieces and then taking the minimum value in conformity with the flaw acceptance standards. The fracture toughness of a material is essentially characterized as being of a statistical nature. Thus, by treating the crack detectability, the accuracy of sizing and the critical fracture toughness as random variables, and by calculating the probability of failure for appropriate fracture modes based on the concept of reliability engineering, hazards can be evaluated quantitatively.

One of the concepts of the probabilistic fracture mechanics approach to evaluating the structural reliability and safety on the basis of a stress-strength (or demand-capability) model [7] is shown in Fig. 1. After fabrication, a structure is usually submitted to inspection by suitable non-destructive tests in order to detect unacceptable flaws. Occurrence

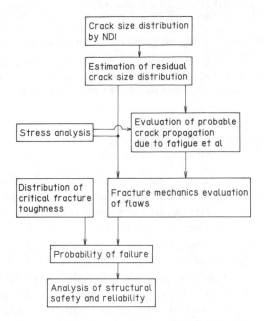

FIG. 1. Probabilistic fracture mechanics approach to structural reliability and safety.

of flaws has been reported in pressure vessels [8–10], offshore structures [11, 12] and steel building construction [13]. These flaws must be repaired before service. The repairs inevitably change the initial defect distribution by screening some of the larger defects. Thus we must estimate the residual crack size distribution which is left in the structure after repairs. The success of a non-destructive test is a function of the crack size and increases following a hyperbolic tangent law with increase in the size. This has been examined experimentally [14–16] and predicted theoretically [17, 18]. On the other hand, the deteriorating effect of the repair weld has been discussed [19, 20], since the residual crack suffers, in some cases, from 'hot straining embrittlement' due to the repair weld. Thus we must consider factors which govern the efficiency and accuracy of non-destructive testing, when the residual crack size distribution is estimated [1, 21, 22].

Depending on operating conditions residual cracks will grow due to fatigue or stress corrosion cracking and so on. Thus the distribution of fracture toughness values of flaws (as demand) can be obtained at the instant of interest by fracture mechanics using the results of stress analysis.

The distribution of the critical fracture toughness of a material (as capability) can also be obtained by performing suitable fracture toughness tests. Scatter in the results of critical fracture toughness tests has been experimentally examined for steels and weldments [21, 23–30], and it has been found to be extremely large in the transition temperature region.

Finally, the probability of failure, P_f, can be obtained as a joint probability of that for the critical fracture toughness x_c being between x_c and $x_c + dx_c$ and that for the fracture toughness x exceeding x_c as

$$P_f = \int_0^\infty \left[\int_{x_c}^\infty g(x)\, dx \right] f(x_c)\, dx_c \tag{1}$$

where $g(x)$ and $f(x_c)$ are the frequency functions of demand and capability, respectively.

Using the concept of failure probability, much analysis has been done concerning other factors such as the reasonable inspection interval [21, 31, 32], parameter sensitivity study [33, 34] and cost benefit analysis [21]. Other practical applications of probabilistic fracture mechanics on other structures such as nuclear power plants, offshore structures, medical equipment, aerospace systems, chemical process plants and bridges are found in recent publications on structural safety and reliability [35].

In the probabilistic fracture mechanics approach, perhaps the greatest problem is the determination of distribution functions of the fracture toughness as both demand and capability. It has been pointed out that the probability of failure differs by a factor of 10^3 depending on the precision of the curve fitting [36].

In this paper, the distribution of the critical fracture toughness of steels is theoretically considered based on experimental results and fractographic observations. Applicability of the beta distribution for the fracture toughness is also discussed.

EXPERIMENTS

Scatter in the critical crack tip opening displacement, δ_c, of a low alloy steel for welded structures (JIS G3106 SM50B) was studied experimentally in the transition temperature region. The plate thickness was 40 mm, and fracture toughness tests on 30 standard-size test pieces 40 mm thick were conducted to obtain δ_c at $-70\,^\circ$C in accordance with the method described in BS5762 [37]. The yield strength at $-70\,^\circ$C was 412 MPa. The stress intensity factor for the fatigue pre-cracking was about $23\cdot3$ MPa$\sqrt{m}$.

The results are shown in Table 1. The δ_c showed a considerable scatter from $0\cdot11$ mm to $0\cdot765$ mm. All data in Table 1 exceed the small scale yield condition [38], but lie far below general yielding [39]. Figure 2 shows a Weibull plot of the data, and fitting seems to be good except for

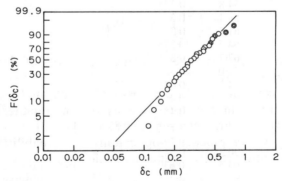

FIG. 2. Weibull plot of critical crack tip displacement δ_c.

the smaller δ_c values. In Fig. 2, the double circular marks denote that the cleavage fracture initiated after a small amount of ductile crack advance (fibrous thumbnail crack), while the others exhibited cleavage fracture only after a certain amount of stretched zone was formed. The shape parameter and the scale parameter determined by the maximum likelihood estimation method are 2·07 and 0·109, respectively.

Figure 3 shows examples of photomicrographs of a fracture surface by means of a scanning electron microscope. The crack initiation site was determined by pursuing the chevron marking inversely to its origin

TABLE 1
Fracture toughness test results

Ordering number	$F(\delta_c)$ (%)	δ_c (mm)	x (mm)	x/δ_c	Fib. thumb crack
1	3·2	0·110	0·14	1·27	
2	6·5	0·126			
3	9·7	0·143	0·10	0·70	
4	12·9	0·151			
5	16·1	0·171			
6	19·4	0·175			
7	22·6	0·201	0·41	2·04	
8	25·8	0·205			
9	29·0	0·221	0·31	1·40	
10	32·3	0·236			
11	35·5	0·252			
12	38·7	0·265			
13	41·9	0·267			
14	45·2	0·288	0·35	1·22	
15	48·4	0·293	0·64	2·18	
16	51·6	0·312			
17	54·8	0·324			
18	58·1	0·330			
19	61·3	0·352			
20	64·5	0·375			
21	67·7	0·375			
22	71·0	0·384			
23	74·2	0·443			
24	77·4	0·454			
25	80·7	0·454	0·76	1·67	observed
26	83·9	0·472	0·85	1·80	
27	87·1	0·501			observed
28	90·3	0·543			
29	93·5	0·639	1·06	1·66	observed
30	96·8	0·765			observed

as shown in (a) and (b) in which the origins are shown by circles. In (c) and (d), which are enlargements of (a) and (b) respectively, subcritical microcracks are shown by arrows. It was not confirmed, however, whether the microcracks were triggered by precipitations, inclusions or

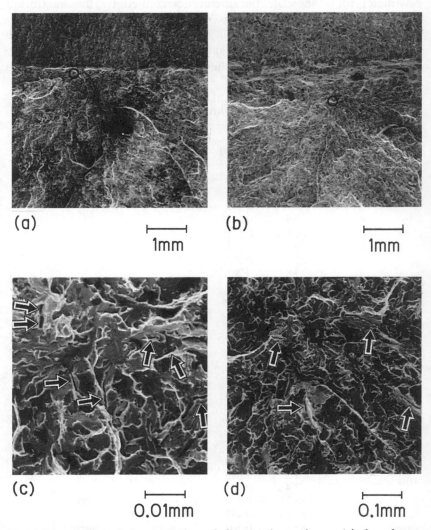

FIG. 3. Examples of fractography of fractured specimen (circles denote initiation site of cleavage fracture): (a) $\delta_c = 0.11$ mm; (b) $\delta_c = 0.639$ mm; (c) and (d) are enlargements of (a) and (b) (arrows indicate subcritical microcracks).

crystallographical inhomogeneities etc. The distance of the crack initiation site, x, from the fatigue crack tip is also listed in Table 1. The distance x is plotted as a function of δ_c in Fig. 4. A good linear relationship was observed even if the data showing ductile crack advance prior to the cleavage fracture (double circles) are included.

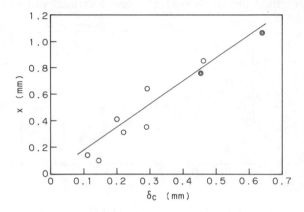

FIG. 4. Relationship between distance of crack initiation site x and δ_c.

DISCUSSION

Statistical Distribution of Fracture Toughness of Steels

McMeeking [40] has computed the maximum principal stress distribution ahead of the crack tip under both the small scale yield and the contained yield condition using finite element calculations. If the stress distribution is plotted as a function of the distance from the undeformed crack tip normalized by the crack tip opening displacement x/δ_t, the form of the stress field is independent of the applied loading. Thus the location where the stress becomes a maximum x_s/δ_t, is constant for a specified material. For example, x_s/δ_t is approximately 3·0, 2·4 and 1·4 for materials having strain hardening exponents 0, 0·1 and 0·2, respectively.

The test results shown in Fig. 4 are replotted in Fig. 5 as x/δ_c versus δ_c. It can be seen from Fig. 5 that the value of x/δ_c, except for the two smallest data, is almost constant, 1·7. These experimental results tell us that the cleavage fracture originates at around the maximum stress location. This seems to coincide well with the prediction by the finite element calculation, since in the calculation a yield strength to Young's modulus ratio, σ_y/E, of 1/300 was used, while in the experiment σ_y/E is about 1/500. This agreement and the observation of subcritical microcracks near the fracture initiation site show that the cleavage fracture will take place when critical microcracks exist within a certain volume in which the stress is larger than a critical value.

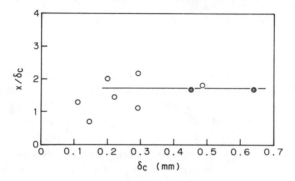

FIG. 5. Relationship between x/δ_c and δ_c.

On the other hand, Hunt and McCartney [41] have developed the probability of defect-initiated failure occurring during the stress increase as

$$P_f = 1 - \exp\left[-V \int_{r_c}^{\infty} h(r)\,\mathrm{d}r \right] \qquad (2)$$

where V is the volume of specimen under consideration, $h(r)$ is the frequency function of defect size which is sparsely distributed in the material, and r_c is the critical size of defect which subsequently develops critical microcracks.

Thus r_c is assumed to satisfy the Griffith crack advance criterion [42]:

$$\sigma_c = [4E(2\gamma_s + \gamma_p)/(1 - v^2)r_c]^{1/2} \qquad (3)$$

where γ_s and γ_p are the surface energy and the plastic work done for crack instability, respectively. According to Eqn. (3), r_c varies as σ varies ahead of the crack tip, and there is a possibility that some of the large defects fail even if the stress is low. However, it is physically reasonable that the probability of fracture being caused by defects in the region of the larger stress is high, since the microcracks are observed only in the vicinity of the maximum stress region.

Thus, assuming that only the vicinity of the maximum stress contributes to fracture [43], a statistical model of the distribution of the critical fracture toughness of steels is derived. The model is shown schematically in Fig. 6. In Fig. 6, σ_{yy} is the stress ahead of the crack tip, x is the distance from the undeformed crack tip and δ_t is the crack opening displacement of the current crack. The effective volume, V_{eff}, involved in the fracture process was evaluated as the volume within which σ_{yy} is larger than a certain value σ_c as follows:

$$V_{eff} = \frac{\pi}{4} x_{eff}^2 \cdot T_{eff} \qquad (4)$$

where x_{eff} is the diameter of a circle whose center is located at x_s (distance from the undeformed crack tip to the location of the maximum stress) and T_{eff} is the effective thickness. Now x_{eff} is taken as Ax_s where A is a proportional constant. As mentioned above, x_s/δ_t is constant for a specified material, so x_{eff} is again proportional to $A\delta_t$. The proportional constant A, strictly speaking, can be obtained by solving the McMeeking stress function with respect to x; namely, x_{eff} is the range where the stress exceeds σ_c

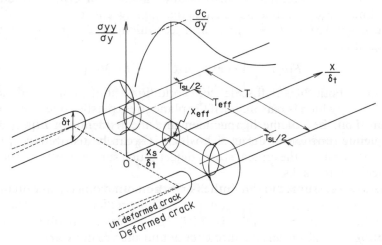

FIG. 6. Schematic representation of stress distribution and effective volume involved in fracture process.

According to observations on the fracture surfaces in the temperature region where the fracture surface consisted of flat fracture and slant fracture, the width of slant fracture T_{sL} increases with increase in the applied load under increasing load test. It has also been reported that T_{sL} is proportional to the plastic zone size under the plane stress condition. The proportional constant is about 0.33 [44] or 1 [45]. Thus T_{eff} is taken as the specimen thickness T_{sL} where the plane strain condition is predominantly satisfied:

$$T_{eff} = T - 0.3BE\delta_t/(1 - v^2)\sigma_y \tag{5}$$

where B is a constant between 0.33 and 1, v is Poisson's ratio and T is the specimen thickness. The relationship between δ_t and crack extension force, G ($= 1.8\sigma_y\delta_t$), derived by Rice [46] was used to obtain Eqn. (5). Substituting V_{eff} for V in Eqn. (2), we get Eqn. (6) for the failure probability:

$$P_f = 1 - \exp\left[-(T - 0.3BE\delta_t/(1 - v^2)\sigma_y)\frac{\pi}{4}A^2\delta_t^2 \int_{r_c}^{\infty} h(r)\,dr \right] \tag{6}$$

If the frequency function of the defect and the species of defect and material are once determined, the integral in Eqn. (6) becomes a constant, C. Therefore the distribution function of critical fracture toughness, $F(\delta_c)$, can be obtained using the new constants A ($= \pi A^2/4$) and B ($= 0.3BE/T(1 - v^2)\sigma_y$), and C which is the result of the integration as

$$F(\delta_c) = 1 - \exp[-ACT(1 - B\delta_c)\delta_c^2] \tag{7}$$

Under both the small scale yield condition and the contained yield condition (which is reasonable if the plane stress plastic zone size is less than about 20% of the ligament length), the term $B\delta_c$ in Eqn. (7) is negligible compared to unity. Thus Eqn. (7) can be reduced as

$$F(\delta_c) = 1 - \exp(-ACT\delta_c^2) \tag{8}$$

Equation (8) is exactly a Weibull distribution function [47] and predicts that the shape parameter is equal to 2. This prediction coincides well with the experimental results (i.e. 2.07). The defect size distribution, $h(r)$, and the critical value, r_c, contribute to the scale parameter.

Prediction of Valid Fracture Toughness from Small Specimen Fracture Toughness

In order to obtain a valid critical fracture toughness, a large specimen is sometimes required. For example, we need a specimen 100 mm thick to obtain the valid critical stress intensity factor, K_{Ic}, of 120 MPa$\sqrt{m}$ for a steel of 600 MPa yield strength.

Thus, development of a method which predicts the valid fracture toughness using small specimens is desired. From the engineering point of view, Landes and Shaffer [29] and Iwadate *et al.* [30] have proposed a method to predict the large specimen fracture toughness using large numbers of small specimens on the basis of the weakest link hypothesis. Namely, the toughness is variable throughout the material, particularly along the crack front. Thus the overall toughness is determined by the lowest toughness along the crack front. Larger specimens contain a larger number of these low toughness regions and the occurrence of a low toughness value is more likely. Smaller specimens, on the contrary, sample less of the variation in toughness and show a great amount of scatter ranging from values near the large specimen toughness to values much greater than the large specimen toughness. Therefore it has been pointed out that the minimum fracture toughness of the small specimens could well coincide with the large specimen fracture toughness. In spite of the agreement between the predicted toughness and the experimentally obtained toughness [29, 30], there are objections concerning the weakest link model. It has been criticized for over-emphasizing the importance of the local value of toughness [48]. It may well be that unstable crack growth is initiated in some larger region of low average toughness. Thus the coincidence of the results could be explained by the additional effect of the loss of constraint in the smaller specimen toughness.

Let us consider a prediction method using the results of Eqns. (7) and (8). Supposing that $\bar{\delta}_{cL}$ and $\bar{\delta}_{cs}$ are the average fracture toughness obtained by the tests using large specimens and small specimens, respectively, we can obtain from Eqn. (8) the ratio of the toughness under the contained yield condition as

$$\frac{\bar{\delta}_{cL}}{\bar{\delta}_{cs}} = \left(\frac{ACT_s}{ACT_L}\right)^{1/2} \frac{\Gamma(1 + \frac{1}{2})}{\Gamma(1 + \frac{1}{2})} = \left(\frac{T_s}{T_L}\right)^{1/2} \tag{9}$$

where T_L and T_s are the specimen thickness of the large and the small specimen and $\Gamma(\ )$ is the gamma function. If the contained yield condition is exceeded, the term $B\delta_c$ in Eqn. (7) is no longer negligible. For that case, we might obtain an appropriate result by using side-grooved small specimens and by substituting the net specimen thickness, T_{net}, instead of T_s in Eqn. (9). Temperature and strain rate dependence of the fracture toughness would be also obtained through these effects on the yield strength.

On the other hand, the same results have been obtained based on the local criterion approach for the cleavage failure. Beremin [49] has analyzed the distribution of the critical stress intensity factor K_{Ic} due to slip-induced cleavage fracture of pressure vessel steel (ASTM A508 Class 3) based on the local criterion approach proposed by Pineau [50]. In the local criterion approach, the macroscopic fracture behavior is modeled in terms of the local fracture criterion. In principle, it is based on the elastic–plastic stress–strain history calculated at the point where fracture takes place in conjunction with the utilization of a micromechanics-based model for a given physical process of fracture. Using the local criterion approach, it has been reported that the distribution of K_{Ic} obeys the Weibull distribution with a shape parameter of 4, and that the product of K_{Ic} and the specimen thickness to the power $\frac{1}{4}$ is constant ($K_{Ic} \cdot T^{1/4}$ = const.) [49]. Wallin et al. [51] and Wallin [52] have also analyzed the scatter of K_{Ic} for carbide cracking induced fracture on the basis of a local criterion and have shown that the distribution again obeys the Weibull distribution with a shape parameter of 4 for such steels as pressure vessel steel (ASTM A508 Class 3) and rotor steels (ASTM A471 and ASTM A470) at a wide range of temperatures. All these results support the predictions made by Eqn. (8).

Application of Beta Distribution Function

It is seen that the data points of smaller toughness, returning to Fig. 2, deviate from the Weibull distribution. This phenomenon has been frequently observed in the transition temperature region and it has been explained by the occurrence of different fracture mechanisms [49] or by the existence of a minimum value fracture toughness below which the cleavage crack propagation becomes impossible [52]. The former reason predicts a bimodal Weibull distribution and the latter recommends the use of a three-parameter Weibull distribution function. It is, however, impossible to distinguish which is the correct reason for the deviation in Fig. 2 simply by the observation on the fractomicrograph. In this paper, an examination of the applicability of the beta distribution function was made. As mentioned in the introduction, the curve fitting has a large effect on the calculation of the failure probability.

The frequency function of the beta distribution is defined as

$$f(x) = \frac{\Gamma(p + q)}{(b - a)^{p+q-1}\Gamma(p)\Gamma(q)}(x - a)^{p-1}(b - x)^{q-1} \qquad (10)$$

where $\Gamma(\ )$ is the gamma function and a, b, p and q are parameters to be determined. As a visual aid, $f(x)$ for the case of $a = 0$ and $b = 1$ is shown in Fig. 7 for various values of the parameters p and q.

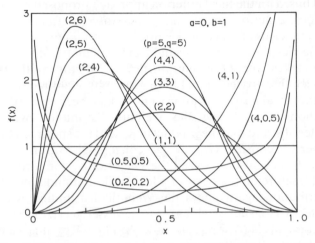

FIG. 7. Beta frequency functions with different parameters.

The mode x_M of $f(x)$ is given by differentiating Eqn. (10) with respect to x and setting $df(x)/dx$ at zero as in

$$x_M = \frac{b(p-1) + a(q-1)}{p + q - 2} \tag{11}$$

The first to the third moments $E(x)$, $E(x^2)$ and $E(x^3)$ are given in Eqns. (12), (13) and (14), respectively:

$$E(x) = \int_a^b xf(x)\,dx = \frac{aq + bp}{p + q} \tag{12}$$

$$E(x^2) = \int_a^b x^2 f(x)\,dx = \frac{(b-a)^2(p+1)p}{(p+q+1)(p-q)} + \frac{2a(b-a)p}{p+q} + a^2 \tag{13}$$

$$E(x^3) = \int_a^b x^3 f(x)\,dx = \frac{(b-a)^3(p+2)(p+1)p}{(p+q+2)(p+q+1)(p+q)}$$

$$+ \frac{3a(b-a)^2(p+1)p}{(p+q+1)(p+q)} + \frac{3a^2(b-a)p}{p+q} + a^3 \tag{14}$$

Now let us obtain the parameters to meet the δ_c distribution given in Table 1. First, the mode of the sample was determined by noting that the data followed the Weibull distribution except for the lower toughness region. Thus, x_M was determined as

$$x_M = \left[\frac{(\alpha - 1)\beta}{\alpha} \right]^{1/\alpha} = 0 \cdot 249 \tag{15}$$

where α and β are the shape parameter and the scale parameter, respectively. The moments of the sample were obtained as

$$\frac{1}{n} \Sigma x_i = 0 \cdot 328 \tag{16}$$

$$\frac{1}{n} \Sigma x_i^2 = 0 \cdot 131 \tag{17}$$

$$\frac{1}{n} \Sigma x_i^3 = 0 \cdot 061 \tag{18}$$

where n is the number of samples and x_i is the data. Equating the above equations to each other, we have four equations such that $(11) = (15)$, $(12) = (16)$, $(13) = (17)$ and $(14) = (18)$. Solving the first two equations as a simultaneous equation, we can get a and b as functions of p and q. Then, inserting a and b into the third equation, a cubic equation can be obtained with respect to q. Solving this cubic equation by the trigonometric method [53], three real roots, q_1, q_2 and q_3, are obtained. Finally, inserting each root into the last equation, we can get an equation with respect to p only. Unfortunately the final equation cannot be handled analytically. Therefore it has been solved by the Newton numerical iteration method [53]. We thus obtain three sets of parameters. Examining each result, the parameters determined are $a = 0 \cdot 061$, $b = 0 \cdot 987$, $p = 2 \cdot 001$ and $q = 5 \cdot 005$.

The beta distribution function, $F_B(\delta_c)$, determined in this manner is shown in Fig. 8 together with the data plot and the Weibull distribution function, $F_W(\delta_c)$. The curve fitting of $F_B(\delta_c)$ seems quite good.

The expected value, μ, the variance, V, and the skewness, S, are obtained, respectively as

$$\mu = E(x) \tag{19}$$

$$V = E(x^2) - E^2(x) \tag{20}$$

$$S = \frac{E(x^3) - 3E(x)E(x^2) + 2E^3(x)}{[E(x^2) - E^2(x)]^{1 \cdot 5}} \tag{21}$$

These values are listed in Table 2 along with the sample values and Weibull values for comparison.

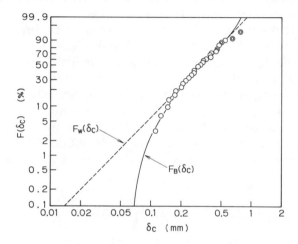

FIG. 8. Data fitting of beta distribution function, $F_B(\delta_c)$, and Weibull distribution function, $F_W(\delta_c)$.

TABLE 2
List of parameters

	Sample		Beta	Weibull
			$a = 0\cdot061$	$\alpha = 2\cdot070$
			$b = 0\cdot987$	
			$p = 2\cdot001$	$\beta = 0\cdot109$
			$q = 5\cdot005$	
$\frac{1}{n}\sum x_i$	$0\cdot328$	$E(x)$	$0\cdot325$	$0\cdot304$
$\frac{1}{n}\sum x_i^2$	$0\cdot131$	$E(x^2)$	$0\cdot128$	$0\cdot116$
$\frac{1}{n}\sum x_i^3$	$0\cdot061$	$E(x^3)$	$0\cdot058$	$0\cdot052$
		x_M	$0\cdot246$	$0\cdot249$
μ	$0\cdot328$		$0\cdot325$	$0\cdot304$
V	$0\cdot023$		$0\cdot022$	$0\cdot024$
S	$0\cdot745$		$0\cdot555$	$0\cdot662$

For the calculation of the failure probability, it is less important to fit the curve exactly in the higher toughness portion compared with the fit in the lower toughness portion. Therefore substitution of the parameter b by a certain constant drawn from experience would not greatly affect the results. This leaves only three parameters still to be determined, and makes the above manipulation considerably easier. Similarly, the parameter a could be reasonably taken as zero to fit the beta distribution to the flaw size distribution.

Now let us compare the minimum value of δ_c determined by the beta distribution with the expected values of the minima for extracted samples from a Weibull distribution. The minimum values are also random variables and show a statistical distribution. The frequency function of the minimum values, $f_m(x)$, can be given as if n samples are extracted:

$$f_m(x) = nf(x)\left[1 - \int_0^x f(x)\,dx\right]^{n-1} \tag{22}$$

Considering only the smaller values of x and assuming that n is larger than 1, Eqn. (22) is approximated as

$$f_m(x) = nf(x)\exp\left[-n\int_0^x f(x)\,dx\right] \tag{23}$$

When the Weibull distribution is employed, the expected value, $E_m(x)$, of $f_m(x)$ is

$$E_m(x) = \int_0^\infty x f_m(x)\,dx = \left(\frac{\beta}{n}\right)^{1/\alpha}\left(1 + \frac{1}{\alpha}\right) \tag{24}$$

where α is the shape parameter and β is the scale parameter. Inserting the experimentally obtained values for the parameters ($\alpha = 2{\cdot}07$ and $\beta = 0{\cdot}109$), and equating $E_m(x)$ to a (= $0{\cdot}061$), the result is $n = 27{\cdot}7$. Thus the minimum value of $0{\cdot}061$ corresponds to the average of minimum values obtained from about 28 test pieces.

According to Wilks [54], the degree of confidence for the quantity of samples to exceed the minimum of results is given approximately as

$$Q^n = 1 - R \tag{25}$$

In Eqn. (25), it is predicted that the percentage, Q, of data from the population would be expected to lie above the minimum with the confidence of R. That is, 92% of the 28 samples are forecast to lie above the minimum value with 90% confidence.

CONCLUSIONS

(1) The statistical distribution of the fracture toughness of steels has been theoretically analyzed on the basis of the observation on the initiation site of the cleavage fracture and the probability of defects existing in the effective volume. The analysis showed that δ_c for steels obeyed a Weibull distribution and that the shape parameter was 2. This prediction coincided well with the experimental results.

(2) Under the contained yield condition, a method was proposed to obtain the valid fracture toughness value from the results of small specimen fracture toughness tests.

(3) The beta distribution function was proposed and analyzed to fit the fracture toughness distribution in order to calculate more accurately the failure probability.

ACKNOWLEDGEMENTS

The author is very grateful to Dr. F. Mudry, of Ecole Nationale Superieure des Mines de Paris, for his helpful discussion on the local criterion theory.

REFERENCES

[1]   N. Urabe, M. Kodama, H. Teraoka, T. Fujita and M. Kawahara, in *Structural Safety and Reliability*, II, IASSAR, 101 (1985).
[2]   N. Urabe, *Nondestruct. Testing J.*, **1**, 187 (1983).
[3]   ASME, *Boiler and Pressure Vessel Code*, Section III, App. G, ASME (1983).
[4]   ASME, *Boiler and Pressure Vessel Code*, Section XI, App. A, ASME (1980).
[5]   BSI, *Published Document 6493*, BSI (1980).
[6]   Japan Weld. Eng. Soc., *WES 2805*, WES (1980).
[7]   A. M. Freudenthal, *Trans. ASCE*, **121**, 1337 (1956).
[8]   P. E. Becher and B. Hansen, *Danish Weld. Inst. Report* (1974).
[9]   E. H. Ruescher and H. C. Graber, *Weld. Res. Council Bull.* (1980).
[10] Y. Maruyama and H. Teraoka, *J. JSNDI*, **31**, 158, 160 (1982).

[11] P. E. L. B. Rodrigues, W. K. Wong and J. H. Rongerson, Offshore Techn. Conf., paper 3693 (1980).

[12] W. K. Wong and J. H. Rongerson, Offshore Techn. Conf., paper 4327 (1982).

[13] T. Nakatsuji, T. Fujimori, M. Kuramochi, K. Satoh and M. Toyoda, *J. JSNDI*, **30**, 191, 257 (1981).

[14] P. F. Packmann, H. S. Pearson, J. S. Owens and G. Young, *J. Mater.*, **4**, 666 (1969).

[15] W. D. Rummel, P. H. Todd, Jr., A. Sandor, A. Freoska and R. A. Ratbke, NASA Contr. Rep. CR-236 (1974).

[16] J. C. Herr and G. L. Marsh, *Mater. Eval.*, Dec., 41 (1978).

[17] M. Ichikawa, *Rel. Eng.*, **9**, 221 (1984).

[18] M. Ichikawa, *Rel. Eng.*, **10**, 175 (1984).

[19] H. Itagaki, M. Toyoda, A. Majima and H. Asada, *J. Soc. Naval Architects, Japan*, **149**, 668 (1981); **152**, 317 (1982).

[20] H. Itagaki, M. Toyoda, Y. Fukugawa, A. Majima and H. Asada, *J. Soc. Naval Architects, Japan*, **150**, 389 (1981).

[21] N. Urabe and A. Yoshitake, *High Press. Techn.*, **19**, 87 (1981).

[22] K. Satoh, M. Toyoda, F. Minami, T. Fujimori and T. Nakatsuji, *J. Japan Weld. Soc.*, **50**, 47 (1981).

[23] T. Kanazawa, H. Itagaki, S. Machida and Y. Kawamoto, *J. Soc. Naval Architects, Japan*, **146**, 444 (1979).

[24] T. Kanazawa, S. Machida and H. Yoshinari, *J. Soc. Naval Architects, Japan*, **150**, 532 (1981).

[25] G. O. Johnston, *Weld. Inst. Res. Rept.*, 106 (1979).

[26] G. O. Johnston, *Weld. Inst. Res. Rept.*, 110 (1980).

[27] G. O. Johnston, in *Probabilistic Fracture Mechanics and Fatigue Methods*, STP 798, ASTM, 42 (1983).

[28] F. Nilsson and B. Ostensson, *Eng. Fract. Mech.*, **10**, 223 (1978).

[29] L. D. Landes and D. H. Shaffer, in *Fracture Mechanics*, STP 700, ASTM, 368 (1980).

[30] T. Iwadate, Y. Tanaka, S. Ono and J. Watanabe, in *Elastic-Plastic Fracture*, STP 803, II, ASTM, 531 (1983).

[31] H. Kitagawa and T. Hisada, *J. High Press. Inst. Japan*, **17**, 270 (1979).

[32] N. Yazdai and P. Albrecht, in *Structural Safety and Reliability*, III, IASSAR, 23 (1985).

[33] M. Shinozuka, *Development of Reliability-Based Aircraft Safety Criteria: An Impact Analysis*, 1, AFFDL-TR-76-31 (1976).

[34] H. Takashima, T. Mimaki and S. Yanagimoto, *J. Soc. Naval Architects, Japan*, **155**, 226 (1984).

[35] IASSAR, *Structural Safety and Reliability*, IASSAR (1985).

[36] A. Bruckner and D. Munz, *Rel. Eng.*, **5**, 139 (1983).

[37] BSI, *BS5762*, BSI (1979).

[38] ASTM, *Annual Book of ASTM Standards*, Part 31, E399, ASTM (1978).

[39] A. P. Green and B. B. Hundy, *J. Mech. Phys. Solids*, **4**, 128 (1956).

[40] R. M. McMeeking, *J. Mech. Phys. Solids*, **25**, 357 (1977).

[41] R. A. Hunt and L. N. McCartney, *Int. J. Fract.*, **15**, 365 (1979).

[42] A. A. Griffith, *Phil. Trans. Roy. Soc.*, **A221**, 163 (1920).

[43]   R. O. Ritchie, J. F. Knott and J. R. Rice, *J. Mech. Phys. Solids,* **21**, 395 (1973).
[44]   K. Nakasa and H. Takei, *Trans. ISIJ,* **18**, 25 (1978).
[45]   D. L. Holt, P. S. Khor and M. O. Lai, *Eng. Fract. Mech.,* **6**, 307 (1974).
[46]   J. R. Rice, in *Proc. 3rd Int. Congr. Fract.,* II, paper I-441 (1973).
[47]   W. Weibull, *J. Appl. Mech.,* **18**, 293 (1951).
[48]   A. Bruckner and D. Munz, *Rel. Eng.,* **18**, 359 (1983),
[49]   F. M. Beremin, *Metall. Trans.,* **14A**, 2277 (1983).
[50]   A. Pineau, in *Proc. 5th Int. Conf. Fract.,* 2, 533 (1981).
[51]   K. Wallin, T. Saario and K. Torroneu, *Met. Sci.,* **18**, 13 (1984).
[52]   K. Wallin, *Eng. Fract. Mech.,* **19**, 1085 (1984).
[53]   G. A. Korn and T. M. Korn, *Mathematical Handbook for Scientists and Engineers,* McGraw-Hill (1961).
[54]   S. S. Wilks, *Ann. Math. Stat.,* **13**, 400 (1942).

Distribution Characteristics of Fatigue Lives and Fatigue Strengths of Ferrous Metals by the Analysis of *P-N* Data in the JSMS Data Base on Fatigue Strength of Metallic Materials

TSUNESHICHI TANAKA, TATSUO SAKAI
Department of Mechanical Engineering, Faculty of Science and Engineering,
Ritsumeikan University, Tojiin-Kitamachi, Kita-ku, Kyoto 603, Japan

and TAKASHI IWAYA
Numazu College of Technology, Ooka, Numazu, Shizuoka 410, Japan

ABSTRACT

A brief explanation is given of the scope and content of the JSMS data base on fatigue strength of metallic materials. This data base was compiled as a joint JSMS project of the Committee on Fatigue and the Committee on Reliability Engineering. A large body of numerical fatigue data and research related information has been stored on a magnetic tape after processing by computer. The *P-N* data (statistical fatigue test data) of ferrous metals were sorted from the data base, and analyzed statistically to determine the basic nature of the distributions of fatigue lives and fatigue strengths of ferrous metals. All the *P-N* data were divided into four classes: two classes of complete samples respectively for smooth specimens (including 'hourglass' type) and for notched specimens, and two classes of censored samples for these types of specimens. In the analysis of the fatigue life distribution, a reduced form of the Weibull distribution, characterized by the shape parameter A and the ratio of the location and the scale parameters, C/B, was introduced to treat the estimated values of A and C/B as those of the samples from the same population regardless of the material and stress level for each class of *P-N* data. Each class of censored samples of the *P-N* data was also used to analyze the distribution characteristics of endurance limits. The basic parameter values characterizing the distributions of fatigue lives and endurance limits were derived from the analysis, and a procedure was proposed for constructing a failure probability surface as a function of the applied stress and the number of cycles, together with the *P-S-N* curves with given failure probabilities.

INTRODUCTION

A data base on the fatigue strength of metallic materials was compiled as one of the joint projects of the JSMS (the Society of Materials Science, Japan) Committees on Fatigue and on Reliability Engineering. Experimental fatigue data from research over the past 20 years were collected on a nationwide scale, and a large amount of data was donated by 88 researchers who had long been studying fatigue problems at academic and research institutes in the public and private sectors. All the data were computer processed and stored on a magnetic tape as an integrated data base. They were also compiled in a data book; this was issued in three volumes by the JSMS in 1982 [1].

The analysis of the data is now being carried out mainly from a statistical viewpoint by the task force members of the Committee on Fatigue, with the aim of the compilation of a handbook on fatigue design and reliability.

In this paper, the scope and content of the data base are first briefly explained to help understanding of the data base itself. The essential features of the distributions of fatigue lives and fatigue strengths of ferrous metals are then discussed through the statistical analysis of the P–N data (the data of statistical fatigue tests using a number of specimens at the respective stress levels) included in the data base.

In order to evaluate the fatigue strength of a given material from a reliable viewpoint, the value of the stress vs. life relation for prescribed failure probabilities (for short P–S–N relations) must be available for reference. This requirement is difficult to attain, however, since the time and cost in constructing such P–S–N relations for every kind of material are formidable, although great effort has been made in this field [2–5].

The problem is divided into two categories: one is to determine an average or central S–N relation; the other is to determine the distribution of lives and strengths relative to this central curve. Therefore, if the latter problem is solved in a general form covering a wide variety of materials with certain standardized test conditions, even if only approximately, the major difficulty will be overcome and the only remaining problem is to find the central S–N relation of each material which can be obtained by conventional fatigue tests. This is the basic idea of the present study; the P–N data are the most useful for the analysis of the distributions of fatigue lives and strengths in the sense described above.

One of the difficulties encountered in this kind of study is the problem of sorting a number of data sets from different sources with various kinds of materials, different specimen shapes and testing conditions. But since the aim of the study is to find a general trend of the distribution, such a variety of sources may actually be advantageous from the practical point of view of aiming at design application. Therefore, all the P-N data were divided into only four classes, categorized either as complete or censored samples or as data from smooth or hourglass type specimens or of notched specimens.

First the three-parameter Weibull distribution was used to analyze the fatigue life distribution, and the parameters were estimated for each set of P-N data. Then the reduced form of this Weibull distribution was introduced to treat the distributions of the shape parameter and the ratio of the location and the scale parameters as if the samples were from a single population in each class of P-N data. The results obtained are comprehensive and indicative of the statistical nature of the fatigue life distribution, including the basic parameter values characterizing the distribution, at least in the sense of a first approximation.

In the latter part of this study, the distribution of the endurance limits was analyzed by applying the normal distribution to the failure probabilities at approximately 10^7 cycles; these were obtained from the P-N data censored at that number of cycles. Basic information concerning the overall trend of the distribution of endurance limits of ferrous metals was obtained from the derived distributions of the sample mean, the sample variance and the coefficient of variation.

Finally, an explanation is given here concerning the method of presentation of a failure probability surface as a function of the applied stress and the number of cycles, together with the P-S-N curves resting on this surface.

SCOPE AND CONTENT OF THE DATA BASE

The JSMS data base on fatigue strength of metallic materials was compiled with the following restrictions regarding the materials and testing conditions in order to avoid excessive diversification.

Materials: ferrous and nonferrous metals, exclusive of welded joints and clad metals.

Type of fatigue test: load or displacement controlled tension–compression, rotating bending, plane bending and torsional fatigue test with constant stress amplitude and mean stress.

Test environments: ordinary atmospheric environment at room temperature; similar environments at controlled temperature, humidity, and pressure (vacuum); and inert and hydrogen gas environments.

Test date: the data are restricted to tests completed after 1961.

Fatigue test data were offered by contributors in one of the following three categories: (i) *S–N*: fatigue test data gathered by ordinary testing methods to obtain a conventional *S–N* relation, with no intention of statistical analysis; (ii) *S–T*: staircase data; fatigue test data obtained by the 'staircase' method; (iii) *P–N*: fatigue test data obtained by using a large number of specimens at one or more stress levels with the intention of statistical analysis.

The following data were also offered to supplement each set of fatigue test data: information on material specifications and processing; lists of chemical composition; records of heat treatments; data on mechanical properties, size, and dimensions of the fatigue specimen together with its preparation and finishing conditions; and fatigue testing conditions, including the load and environment.

All these data were grouped into 'fundamental sets' and stored in the data base according to a key code. A fundamental set of data provides full information on a test series carried out under the same testing conditions with the same material and the same specimen.

The total number of sets contained in the data base is 2426. The numbers of sets of *S–N*, *S–T*, and *P–N* data respectively for ferrous and nonferrous metals are listed in Table 1. The data sets for ferrous metals occupy 89% of the total. The number of data points corresponding to the number of specimens used in fatigue tests is 35 614 in total, and those for *S–N*, *S–T*, and *P–N* data are listed in Table 2. These sets are also grouped according to loading types, specimen forms, and test environments respectively, as shown in Tables 3–5. In these tables, ferrous metals include carbon and alloy steels for machine structural use, rolled steels

TABLE 1
Number of sets for three types of data

Metals	SN	ST	PN	Total
Ferrous	1 934	112	111	2 157
Nonferrous	226	4	39	269
Total	2 160	116	150	2 426

for general structures, boiler steels, heat resisting steels, tool steels, spring steels, stainless steels, steel forgings, iron and steel castings, and other commercial base steels, including high tension steels. Nonferrous metals include aluminium and aluminium alloys, aluminium alloy castings, and copper and copper alloys. Certain of the codes referring to loading types and specimen forms appearing in Tables 3 and 4 will be used in this paper.

TABLE 2
Number of data points for three types of data

Metals	SN	ST	PN	Total
Ferrous	18 006	1 989	9 287	29 282
Nonferrous	2 986	119	3 227	6 332
Total	20 992	2 108	12 514	35 614

TABLE 3
Number of sets grouped by loading types

Metals	AX	RB	RC	BI	BO	BB	TW	Total
Ferrous	396	1 295	149	56	146	37	78	2 157
Nonferrous	106	104	6	1	27	14	11	269
Total	502	1 399	155	57	173	51 ·	89	2 426

AX, axial loading; RB, uniform rotating bending; RC, cantilever rotating bending; BI, in-plane bending of plate; BO, out-of-plane bending of plate; BB, bending of round bar; TW, torsion.

TABLE 4
Number of sets grouped by specimen forms

Metals	SM	RD	NT	OT	Total
Ferrous	1 063	384	635	75	2 157
Nonferrous	166	20	83	0	269
Total	1 229	404	718	75	2 426

SM, smooth specimen; RD, hourglass type specimen (round bar or plate); NT, notched specimen; OT, others.

TABLE 5
Number of sets grouped by test environment

Metals	AE	AR	N2	OT	Total
Ferrous	2 118	6	13	20	2 157
Nonferrous	262	2	5	0	269
Total	2 380	8	18	20	2 426

AE, room air; AR, argon gas; N2, nitrogen gas; OT, other atmosphere.

EVALUATION OF THE FATIGUE LIFE DISTRIBUTION OF FERROUS METALS

Analytic Procedure

The purpose of this study is to determine the distribution chracteristics of fatigue lives and strengths relative to a central S–N relation; the most useful information for this purpose is stored in the P–N data, whose sizes are shown in Tables 1 and 2 in different forms. In this section, the fatigue life distribution is discussed through the analysis of the complete and censored samples of the P–N data. This procedure is outlined first.

The three-parameter Weibull distribution was used to approximate empirical distributions of fatigue life data at the respective stress levels included in each fundamental set of P–N data; this was used because this type of distribution is more advantageous in discussing the lower limit of the distribution than the conventional log-normal distribution frequently used in similar analyses.

The three-parameter Weibull distribution is expressed as

$$F_w(N) = 1 - \exp\{-[(N - C)/B]^A\} \tag{1}$$

where A, B and C are the shape, scale and location parameters, and N is the fatigue life. By specifying

$$Y = \ln\ln\{1 - F_w(N)\}^{-1} \quad \text{and} \quad X = \ln(N - C) \tag{2}$$

we obtain the linear expression

$$Y = A(X - \ln B) \tag{3}$$

Figure 1 illustrates an experimental fatigue life distribution approximated by Eqn. (1) together with the linear relation of Eqn. (3) on Weibull probability paper. Making use of Eqn. (3), the two parameters A and B

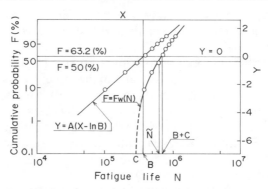

FIG. 1. Illustration of the three-parameter Weibull distribution.

can be estimated by the linear regression method when C is properly estimated. The correlation coefficient method developed by the authors [6–8] was used to estimate the three parameters for each set of fatigue life data (for each sample) at each stress level. In this method, C is estimated by a computer-aided trial and error technique as a value where the correlation coefficient of the points located along the straight line in Fig. 1 reaches a maximum.

According to previous studies on carbon and alloy steels, the failure probability vs. life curves (P-N curves) plotted on probability paper, as shown in Fig. 1, are approximately parallel to each other when the applied stresses are relatively high [2–5]. This means that the parameter A and the ratio C/B are almost independent of the stress and that they are approximately constant. If this is the case, Eqn. (1) is reduced to the form

$$F_\xi(\xi) = 1 - \exp\{-(\xi - \gamma)^A\} \tag{4}$$

where $\xi = N/B$ and $\gamma = C/B$. All the P-N curves at different stress levels can then be represented by a single curve, expressed by Eqn. (4). Since $B + C$ is close to the central value $\tilde{N}$, as shown in Fig. 1, $C/(B + C) = \gamma/(1 + \gamma)$ gives an approximate value of the ratio $C/\tilde{N}$, i.e. the ratio of the lower limit to the median value of the distribution.

Now, if B is known for a sample of size n, and when N_1 is the shortest fatigue life in the sample, then $\xi_1 = N_1/B$ is an estimate of γ, and the distribution of ξ_1 is given by

$$\Phi_n(\xi_1) = 1 - \{1 - F_\xi(\xi_1)\}^n$$
$$= 1 - \exp\{-(\xi_1 - \gamma)^{nA}\} \tag{5}$$

Consequently, ξ_1 obeys the same type of distribution as specified by Eqn. (4) with the shape parameter nA, and γ is given by the lower limit of the distribution of ξ_1.

According to the authors' previous investigations [6–8], this correlation coefficient method gives good estimates of the three parameters A, B and C when the sample size n is large (i.e. where $n \geqslant 30$), but when n becomes small ($n < 30$) the estimation of C rapidly becomes erroneous. Consequently, in the analysis of the P–N data in subsequent sections, the parameters A and B were estimated by the correlation coefficient method and $\gamma = C/B$ was estimated as the lowest value of the distribution of ξ_1 given by Eqn. (5).

For censored samples (data sets containing those derived from run-out specimens), only the data of finite lives were assumed to make a complete sample at each stress level and analyzed in a similar way.

Estimation of The Shape Parameter

Complete Samples for Smooth and Hourglass Type Specimens
The complete samples obtained for the smooth (SM) and hourglass type (RD) specimens were analyzed first. The distribution of the estimated shape parameters, A, is presented on log-normal probability paper in Fig. 2. This class of P–N data covers the test data of a variety of materials under different types of load, as indicated in Fig. 2. The types of load are indicated by the same codes as in Table 3, and the materials used are indicated by JIS symbols. Codes FC, AC, and Q&T in parentheses refer to furnace cooling, air cooling, and quenching and tempering, respectively. High temperature test data on S10C steel and data on tufftrided S45C steel are also included. Each figure in the three columns indicates the number of test series, the number of stress levels, and the number of specimens included in each item of the load and the material. The total number of points in the diagram is 271, equal to the total number of stress levels; the set of data at each stress level yields one estimate of A.

Since the sample size of each set of data has a wide variety, ranging from 5 to 77, the plot was made by dividing the sample sizes into 6 groups, as indicated in Fig. 2, in order to ascertain the dependence of distribution on sample size. It was observed that each distribution of A is approximately linear on log-normal probability paper, and the distribution range becomes narrower with increases in sample size. The median of the distribution is $\tilde{A} \simeq 1.48$, which was determined as an

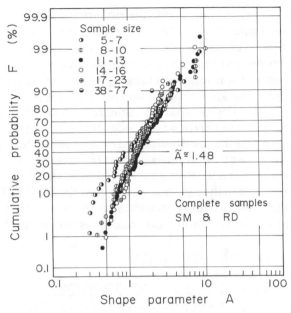

FIG. 2. Distribution of the estimated shape parameter A; complete samples for smooth (SM) and hourglass type (RD) specimens of carbon and alloy steels.

Load	Material	Series	Levels	Specimens
	S10C–S45C (FC, AC, Q&T)	27	100	1 317
	SS41, SC49, SF50, SWRH62A	5	14	237
RB	SCM435, SNCM439, SK5	14	85	1 069
RC	SUS316, SUS403	3	12	118
	S10C (high temp.)	3	10	150
	S45C (tufftrided)	1	2	40
BO	S10C, S50C, SM50A, SUJ2	10	30	382
BB	SUJ2	7	7	35
AX	S15C, S35C	2	3	35
TW	S45C, SUJ2	8	8	61
		80	271	3 444

average of the 50% probability values of the distributions for respective sample sizes. The most favorable consequence induced from Fig. 2 may be that all the estimates of A are from the samples belonging to the same population, with a population parameter $\tilde{A} \simeq 1\cdot48$; even if this is confirmed only as a first approximation, a great improvement will be made in the analysis of fatigue life distribution of ferrous metals.

In order to examine this hypothesis, the results of the similar analysis for the more limited cases of loads and materials are presented in Fig. 3; the diagram to the left (a) shows the distribution of A when (1) the material is limited to carbon steels ranging from S10C to S45C and (2) the load is limited to uniform (RB) and cantilever type (RC) rotating bending load; the diagram to the right (b) shows the results for quenched and tempered carbon steel S45C tested under RB load. It is interesting to see that the distribution characteristics of the estimated A values are nearly identical in the diagrams (a) and (b), and they have almost the same characteristics as that in Fig. 2; the median of the distribution $\tilde{A} \simeq 1{\cdot}46$ in the case of the carbon steels (S10C to S45C), and $\tilde{A} \simeq 1{\cdot}47$ for the quenched and tempered S45C steel. It has been inferred therefore that the distribution of A values, as shown in Fig. 2, is a

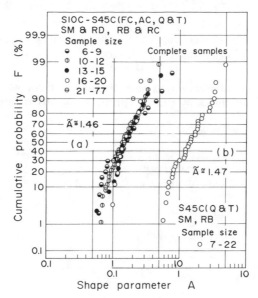

FIG. 3. Distribution of the estimated shape parameter A; complete samples for smooth (SM) and hourglass type (RD) specimens of carbon steels.

Load	Material	Series	Levels	Specimens
RB&RC	S10C–S45C (FC, AC, Q&T)	27	100	1 317
RB	S45C (Q&T)	7	35	410

common feature observed for a wide variety of load conditions and materials for complete samples of SM and RD specimens.

An analytical confirmation of the foregoing consequence concerning the single population behind the distribution of the estimated values of A would be to compare the distribution of A shown in Fig. 2 for each group of a particular sample size with the theoretical distribution for a corresponding size. Using the same correlation coefficient method, the latter distribution can be created by plotting the distribution of the values of A obtained from a large number of sets of data for given sample size taken out of the population having definite parameter values. Such theoretical distributions obtained by a computer-aided Monte Carlo simulation technique [9] are shown in Fig. 4 for sample sizes from 7 to 50. It was assumed that the population has the distribution function given by Eqn. (4), and the parameters were determined as $A = 1 \cdot 5$ and $\gamma = 0 \cdot 1$. The value of γ was determined as described in a later section of this paper. For paper sample size, the parameter estimations were repeatedly carried out on 1000 sets of data which were randomly selected based on the given distribution function.

It was observed that the distribution characteristics given in Fig. 2 are much the same as those in Fig. 4, although the distribution in Fig. 2

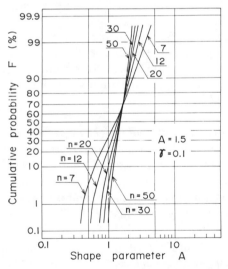

FIG. 4. Theoretical distribution of the sample values of A by a Monte Carlo technique.

tends to spread toward larger values of A compared to that in Fig. 4. However, this deviation of A toward the larger side is allowable from the viewpoint of reliability, since the large value of A implies a small scatter of the fatigue life data. Therefore, the foregoing consequence concerning the single population is acceptable as a reasonable approximation.

Complete Samples for Notched Specimens

The distribution of the values of A estimated from the complete samples for notched (NT) specimens is presented in Fig. 5. The types of both load and materials used in this analysis are listed in the diagram, together with the numbers of test series, stress levels, and specimens belonging to each item. The distribution of A is plotted for each of the three groups of sample size.

It was observed here as well that the distribution is approximately linear on log-normal probability paper, and is similar to that of Fig. 2, although the total number of points is only about half that of the previous case. The distribution median is $\tilde{A} \simeq 2 \cdot 13$ in Fig. 5, which was obtained as an average of the distribution medians for three groups of sample size.

If we accept that the distribution of the values of A indicated in Fig. 5 is a common feature of the shape parameter estimated from complete samples of given sizes taken out of a single population for notched specimens, then the shape parameter A of this population is larger than the corresponding value for smooth and hourglass type specimens; this is clear from a comparison of the values of $\tilde{A}$ in both cases. This implies that the scatter of fatigue life is less for notched specimens than for smooth and hourglass type specimens, as has been pointed out previously [10, 11].

Censored Samples for Smooth and Notched Specimens

Censored samples, truncated at about 10^7 cycles, occur at stress levels in the neighborhood of the endurance limit. At a stress in this neighborhood, if k specimens failed within 10^7 cycles among n specimens, k/n gives the failure probability at 10^7 cycles and is a function of the applied stress; lower stress levels give smaller values of k/n. The estimation of the Weibull parameters for censored fatigue life data was made using the data for k specimens, assuming them to be a complete sample, as described above in the section on analysis procedure.

The distribution of the estimated shape parameter A for smooth and hourglass type specimens is shown in Fig. 6 (a) and (b), and for notched

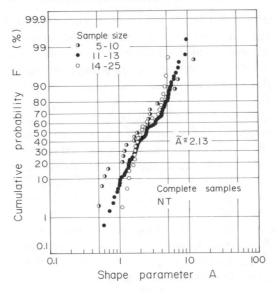

FIG. 5. Distribution of the estimated shape parameter A; complete samples for notched (NT) specimens of carbon and alloy steels.

Load	Material	Series	Levels	Specimens
RB	S45C, S55C (FC, Q&T)	8	40	431
	SCM435, SNCM439, SK5	9	63	802
RC	SUS403	1	9	115
	S35C (press fitted)	1	3	39
BO	SPCC, SM50A	3	17	217
		22	132	1 604

specimens in Fig. 6 (c). The types of load and materials relevant to these classes of *P-N* data are listed in the diagram. The figures in the 'specimens' column indicate the number of failed specimens, with the sum of the failed and run-out specimens shown in parentheses.

In order to see the possible effect of the value of k/n or stress level on the distribution of A for SM and RD specimens, respective distributions for the higher and lower values of k/n separated arbitrarily at $k/n = 0.7$ are shown in diagram (a) of Fig. 6; diagram (b) shows the distribution when they are pooled together. The difference of sample size is ignored in these diagrams, since the number of estimates is not large enough compared to those obtained for the complete samples. It was observed

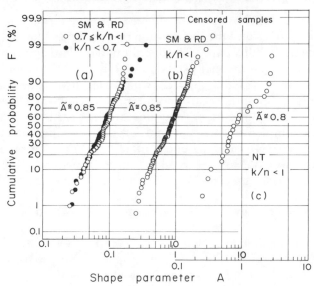

FIG. 6. Distribution of the estimated shape parameter A; censored samples for smooth (SM) and hourglass type (RD) specimens and notched (NT) specimens of carbon and alloy steels.

Load	Material	Series	Levels	Specimens	
Smooth and hourglass type specimens:					
RB RC	S10C–S45C (FC, AC, Q&T)	20	43	587	(779)
	SS41, SWRH62A	3	4	40	(56)
	SCM435, SCM440, SNCM439, SK5	15	32	316	(478)
	SUS316, SUS403	3	4	33	(54)
	S10C (high temp.)	2	3	33	(42)
BO	S10C, S50C, SM50A	4	7	87	(115)
TW	S45C	2	5	32	(52)
		49	98	1 128	(1 576)
Notched specimens:					
RB RC	S10C, S45C, S55C (FC, Q&T)	7	9	71	(121)
	SCM435, SNCM439, SK5	8	9	80	(118)
	SUS403	1	2	20	(26)
	S35C (press fitted)	1	1	8	(13)
BO	SM50A	2	2	17	(26)
		19	23	196	(304)

that the value of k/n has no influence on the distribution of the estimated A values and the distribution itself is again almost linear on log-normal probability paper. The distribution median is approximately $\tilde{A} = 0.85$, which is distinctly less than the value obtained for the complete samples. This fact coincides with the results obtained previously by the authors for certain carbon and alloy steels [3–5].

For the notched specimens (diagram (c)), the distribution characteristics of the sample values of A are not very clear because the number of points is only 23. However, a trend similar to that found for the SM and RD specimens was also observed here, and the median $\tilde{A} = 0.8$ is also distinctly less than the value obtained for the complete samples.

Estimation of The Lower Limit of Fatigue Life Distribution

Complete Samples for Smooth and Hourglass Type Specimens
In the previous section, the distribution characteristics of the estimated shape parameter were analyzed by grouping all the P–N data into four classes — the classes of complete and censored samples for both the SM/RD specimens and NT specimens — and the analysis was carried out on the assumption that Eqn. (4) holds in each class of P–N data. In this section, based on the same assumption, the discussion will concentrate on the lower limit of the fatigue life distribution, beginning with the complete samples for the SM and RD specimens.

Since the lower limit, $\gamma = C/B$, in Eqn. (4) is estimated as the lowest value of the distribution of $\xi_1 = N_1/B$ according to Eqn. (5), the cumulative frequency of N_1/B is plotted on Weibull probability paper for each group of sample size in Fig. 7. The types of load and materials used here are listed as before. The single item of S45C (tufftrided) that appeared in the corresponding list with Fig. 2 is excluded in Fig. 7, since this sample contains some extremely short fatigue life data relative to the average life. Six other data values in the samples whose items are included in Fig. 2 have also been excluded here in plotting the data in Fig. 7. Ignoring these data, the respective distributions have a trend converging approximately to the same lower limit $\gamma = C/B \simeq 0.075$. Therefore, if this value of γ is adopted as the lower limit of the assumed population distribution given by Eqn. (4), there is a risk of $6/3404 = 1.8 \times 10^{-3}$, since six specimens among 3404 gave the still smaller value N_1/B than the above value of γ.

In order to examine the universality of this result obtained for a variety of materials and different types of load, a similar distribution of

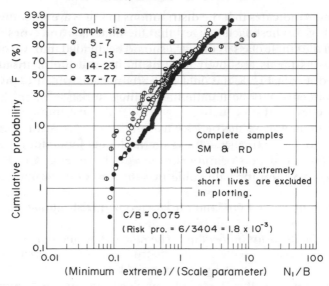

FIG. 7. Distribution of the ratio N_1/B; complete samples for smooth (SM) and hourglass type (RD) specimens of carbon and alloy steels.

Load	Material	Series	Levels	Specimens
	S10C–S45C (FC, AC, Q&T)	27	100	1 317
RB	SS41, SC49, SF50, SWRH62A	5	14	237
RC	SCM435, SNCM439, SK5	14	85	1 069
	SUS316, SUS403	3	12	118
	S10C (high temp.)	3	10	150
BO	S10C, S50C, SM50A, SUJ2	10	30	382
BB	SUJ2	7	70	35
AX	S15C, S35C	2	3	35
TW	S45C, SUJ2	8	8	61
		79	269	3 404

N_1/B is presented in Fig. 8 for the limited set of items identical to those in Fig. 3; i.e. Fig. 8(a) relates to carbon steels from S10C to S45C with FC, AC, and Q&T treatments, and Fig. 8(b) shows only the case of S45C steel with Q&T treatment. It is observed that the distribution characteristics are much the same in both diagrams (a) and (b), though in (b) a single distribution is shown regardless of sample size, and the lower limit of the distribution is estimated as approximately $\gamma = C/B \simeq 0.2$ for both cases if the point near $N_1/B = 0.1$ is ignored. This value of γ is different from the previous one obtained from Fig. 7, and comparison of

Figs. 7 and 8 indicates that the distributions in Fig. 8 are smoother than those in Fig. 7. These facts suggest that the materials and types of load pooled in the single class as in Fig. 7 should be divided into two or more subclasses as far as the lower limit of the fatigue life distribution is concerned. From Fig. 8, the data for carbon steels with FC, AC, and Q&T treatments may be handled as a single subclass when they are obtained under RB and RC loads.

Figure 9 shows the theoretical distributions of N_1/B for respective sample sizes, n, obtained by the same Monte Carlo technique as that employed in Fig. 4. The location parameter of the population was fixed at $\gamma = 0\cdot1$ as an appropriate value by referring to the corresponding lower limits estimated in Figs. 7 and 8. A better agreement between the distribution trends of Fig. 9 and Fig. 8(a) confirms the above discussion concerning the subclass comprising the data for all carbon steels from S10C to S45C. Here attention should be paid to the fact that, in Fig. 9, all

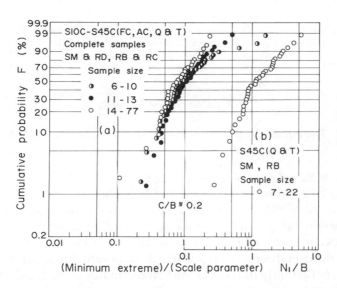

FIG. 8. Distribution of the ratio N_1/B; complete samples for smooth (SM) and hourglass type (RD) specimens of carbon steels.

Load	Material	Series	Levels	Specimens
RB&RC	S10C–S45C (FC, AC, Q&T)	27	100	1 317
RB	S45C (Q&T)	7	35	410

the theoretical distributions of different sample sizes have lower limits less than the previously fixed location parameter of the population, $\gamma = 0.1$. This is certainly due to the employment of the sample value of the scale parameter B for the computation of N_1/B instead of the population value $B = 1$. Consequently, the location parameter, γ, estimated as the lower limit of the distribution of N_1/B is always on the conservative side.

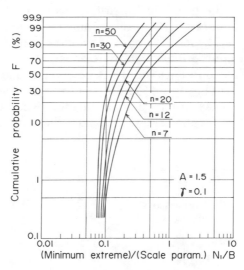

FIG. 9. Theoretical distribution of the ratio N_1/B by a Monte Carlo technique.

Complete Samples for Notched Specimens

Figure 10 shows the distribution of N_1/B obtained from the complete samples of notched specimens. The types of load and materials included in this class are the same as those in Fig. 5, but the sample size is divided into only two groups, rather arbitrarily in this case, in order to reveal the overall trend of the distribution. Several inflection points are found, particularly in the distribution plotted by solid circles, and the curve is no longer smooth. Hence, it seems better to divide this class into several subclasses with respect to the materials, types of load, and notch forms if necessary. Nevertheless, the lower limits of the distributions for two groups of sample size almost coincide with each other and indicate the value $\gamma = C/B \simeq 0.12$, though one sample of the data is excluded in this presentation because of its extremely small value of N_1/B.

Therefore, if this value of C/B is adopted as the lower limit for this class of data, at least as a first approximation, there may be a risk of $1/1604 \simeq 6 \times 10^{-4}$.

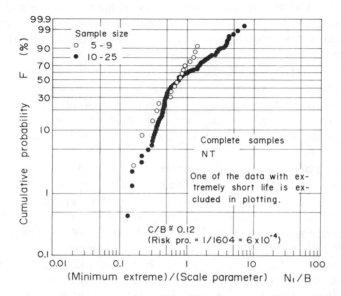

FIG. 10. Distribution of the ratio N_1/B; complete samples for notched (NT) specimens of carbon and alloy steels.

Load	Material	Series	Levels	Specimens
	S45C, S55C (FC, Q&T)	8	40	431
RB	SCM435, SNCM439, SK5	9	63	802
RC	SUS403	1	9	115
	S35C (press fitted)	1	3	39
BO	SPCC, SM50A	3	17	217
		22	132	1 604

Censored Samples for Smooth and Hourglass Type Specimens and for Notched Specimens

For the censored samples, the two classes of data for smooth and hourglass type specimens and notched specimens are pooled together in discussing the lower limit of their life distribution. The reason is that the distribution characteristics of the estimated shape parameters A

were approximately the same for the two classes with respect to linearity on log-normal probability paper, as were the medians, $\tilde{A}$, as shown in Fig. 6.

Figure 11 shows the distribution of N_1/B for these pooled censored samples: diagram (a) shows the respective distributions for the higher and lower values of k/n separated by $k/n = 0.7$, while diagram (b) shows the distribution for the entire range of $k/n < 1$; no distinction was made for sample size. No significant difference was observed between the two distributions plotted here by the open and solid circles, and they coincide therefore with the single distribution in diagram (b). However, the distribution itself is somewhat different from those of the complete samples shown in Figs. 7–9, in that it has a 'tail' extending to an extremely small value of N_1/B.

A value of $\gamma = C/B = 0.01$ is suggested in Fig. 11 as a possible lower limit, simply because the plotted values of N_1/B are all larger than 0.01. This value is, however, less by one order of magnitude than those estimated previously for the complete samples. Another choice is to ignore the four points located along the lower tail of the distribution to make the distribution itself smoother. By such a modification, the lower limit certainly becomes more pronounced and will give the value $\gamma = C/B \simeq 0.05$, which is still less than the value expected at stress levels near the endurance limit. Concerning this point, a survey of the data and the respective values of N_1/B suggested that the denominator B, the sample value of the scale parameter, varies in a wide range due to the reduced accuracy of estimation resulting from the use of data from a relatively small sample size; as mentioned above, only k data of failed specimens among n in total were used for the estimation.

Therefore, as another approach to find a more realistic value of the lower limit of the life distribution, the distribution of the minimum extreme (the shortest life of each set of data) N_1 is plotted in Fig. 12, instead of N_1/B, since N_1 is an estimate of the location parameter, C, in Eqn. (1), and the lower limit of the distribution of N_1 is a much better estimate of C. In plotting such a distribution, it must be assumed that all the censored data are from the same population, having in common certain constant values of the parameters A, B and C, though admittedly this is a rather bold speculation.

In Fig. 12(a), the distributions of N_1 for two different ranges of k/n are presented, and in Fig. 12(b) the distribution for the entire range of k/n is shown. The included items are the same as those in Fig. 11. It can be observed in both diagrams that the lower limit of the distribution is more clearly defined and has a value of approximately 10^5. Accordingly, the location parameter for the censored samples can be assumed to be $C \simeq 10^5$.

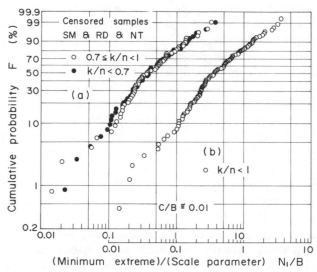

FIG. 11. Distribution of the ratio N_1/B; censored samples for smooth (SM), hourglass type (RD) and notched (NT) specimens of carbon and alloy steels.

Load	Material	Series	Levels	Specimens
	Smooth and hourglass type specimens:			
RB RC	S10C–S45C (FC, AC, Q&T)	20	43	779
	SS41, SWRH62A	3	4	56
	SCM435, SCM440, SNCM439, SK5	15	32	478
	SUS316, SUS403	3	4	54
	S10C (high temp.)	2	3	42
BO	S10C, S50C, SM50A	4	7	115
TW	S45C	2	5	52
		49	98	1 576
	Notched specimens:			
RB RC	S10C, S45C, S55C (FC, Q&T)	7	9	121
	SCM435, SNCM439, SK5	8	9	118
	SUS403	1	2	26
	S35C (press fitted)	1	1	13
BO	SM50A	2	2	26
		19	23	304

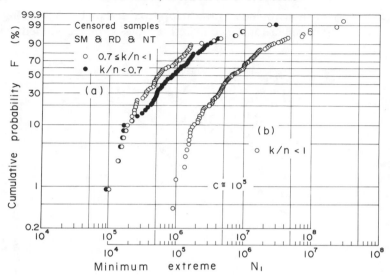

FIG. 12. Distribution of the minimum extreme N_1; censored samples for smooth (SM), hourglass type (RD) and notched (NT) specimens of carbon and alloy steels.

EVALUATION OF THE ENDURANCE LIMIT DISTRIBUTION OF FERROUS METALS

Analytic Procedure

When it is assumed that every $S-N$ curve becomes horizontal before 10^7 cycles for ferrous metals, then if k is the number of specimens failed before 10^7 cycles among n specimens ($k < n$) at a given stress level, the value of k/n gives the failure probability at this stress level. Also, if $p = k/n$ is plotted against the applied stress, the diagram thus obtained represents the distribution of strengths at 10^7 cycles, i.e. the distribution of endurance limits.

The value k/n is obtained at respective stress levels from the censored samples of $P-N$ data, which was already used in the previous discussion in Figs. 6, 11 and 12. However, in order to plot the endurance limit distribution of a given material, one must have the values of k/n at least at two stress levels. Therefore, among the censored samples treated above, some were excluded, and those samples that were obtained under rotating bending (RB and RC) and out-of-plane bending (BO) loads have been analyzed here. The contents of this class of censored *P-N* data are indicated in Table 6; (a) shows the smooth (SM) and hourglass type (RD) specimens and (b) the notched (NT) specimens. Since each test series has *P-N* data at several stress levels obtained under the same test conditions for a certain material, the distribution of endurance limits is plotted for every test series here. However, when $k/n = 0$ the value of p cannot be plotted on probability paper. This case arises when all the specimens run out 10^7 cycles perhaps at the lowest stress level included in a test series. An analogous case occurs when $k/n = 1$ at stress levels higher than those yielding censored samples of a test series; in this case the data belong to the complete samples. These two cases give us certain information about the terminal points of the distribution, but the corresponding points go to upper and lower infinities when they are plotted on standard probability paper. Therefore these cases are omitted in the following analysis of endurance limit distribution.

Figure 13 is an example of the distribution of endurance limits

TABLE 6
Details of the class of *P-N* data used for the analysis of the endurance limit distribution

Load	Material	Series	Levels	Specimens
(a) SM and RD specimens				
	S10C–S45C (AC, FC, WC, Q&T)	15	79	1 253
RB	SS41	1	3	47
RC	SCM435 (AC, Q&T), SCM440, SNCM439, SK5 (FC, Q&T)	13	61	867
	SUS316, SUS403	1	4	52
BO	S10C (FC), S45C, SM50A	4	20	271
Total		34	167	2 490
(b) NT specimens				
RB	S45C (Q&T), S50C (FC)	4	16	179
	SCM435 (Q&T), SNCM439 (Q&T), SK5 (AC, FC)	7	25	317
RC	SUS403 (Q&T)	1	4	53
Total		12	45	549

RB, uniform rotating bending; RC, cantilever rotating bending; BO, out-of-plane bending of plate; SM, smooth specimen; RD, hourglass type specimen (round bar or plate); NT, notched specimen.

plotted on normal probability paper. The top and bottom points with arrows indicate the cases when $k/n = 1$ and 0, respectively. The four points in the range $0 < k/n < 1$ were used to determine the indicated regression line (the weighted Probit method [12, 13] was used for the regression analysis). This method can take the sample size n into account and put larger weight on the points near 50% probability, and less weight on the points far from the center.

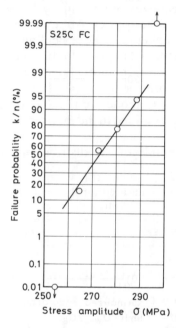

FIG. 13. Example of the distribution of endurance limit.

It was observed that the distribution of the endurance limit is well approximated by the normal distribution, as far as the four points in the range $0 < k/n < 1$ are concerned.

The mean value of the endurance limit, σ_{w50}, and the standard deviation, S, were read from the regression line for each test series. The coefficient of variation was then calculated as S/σ_{w50}.

Evaluation of the Distribution Characteristics of Endurance Limit

Smooth and Hourglass Type Specimens

First, the endurance ratio, defined by $R_w = \sigma_{w50}/\sigma_B$, was calculated for 34 test series using the ultimate tensile strength σ_B of the materials in the case of SM and RD specimens, and the distribution was plotted on normal probability paper, as shown in Fig. 14.

It is interesting to note that the endurance ratio obtained under rotating and plane bending loads approximately follows a normal distribution for the ferrous metals listed in Table 6, with a median value R_{w50} of 0·50 and a standard deviation S_{Rw} of 0·05.

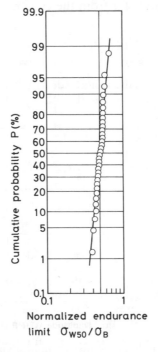

FIG. 14. Distribution of the endurance ratio R_w; smooth and hourglass specimens.

Then, in order to see their dependence on the fatigue strength of material, the standard deviation S and the coefficient of variation v obtained for each test series were plotted against σ_{w50} in Figs. 15 and 16. It was observed that the effect of the strength σ_{w50} is small compared to the scatter bands of S and v. In Fig. 15, however, the upper and lower bounds of the scatter band increase slightly with an increase of σ_{w50}. In Fig. 16, on the other hand, the upper bound decreases with an increase of σ_{w50}, but the lower bound remains almost constant. Comparing the two diagrams, the coefficient of variation v seems to be less dependent on the fatigue strength of material than does the standard deviation S.

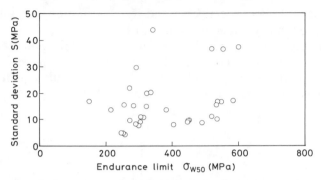

FIG. 15. Correlation between the standard deviation S and the mean endurance limit σ_{w50}; smooth and hourglass specimens.

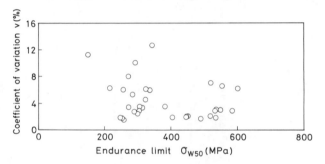

FIG. 16. Correlation between the coefficient of variation v and the mean endurance limit σ_{w50}; smooth and hourglass specimens.

Similar trends of S and v can be observed when they are plotted against the ultimate strength σ_B, as is easily understood from the strong correlation between σ_B and σ_{w50}, shown in Fig. 14.

From the viewpoint of design engineering, it is important to have basic information concerning the distribution characteristics of S and v, because such information can provide risk or safety data when a certain value of S or v is adopted to determine the allowable design stress level. Figures 17 and 18 show such a distribution on log-normal probability paper. The slight influence of the mean strength σ_{w50} is ignored in these figures. Figure 17 shows that the distribution of S is approximated by a straight line, and that S therefore obeys log-normal distribution. Its median is approximately $\tilde{S} = 13\cdot2$ MPa. In Fig. 18 the distribution of v is also approximated by log-normal distribution, and from the straight line approximation the median v is obtained as $\tilde{v} = 3\cdot7\%$.

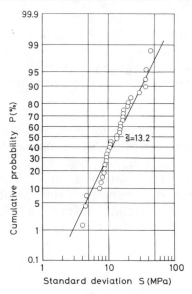

FIG. 17. Distribution of the standard deviation S; smooth and hourglass specimens.

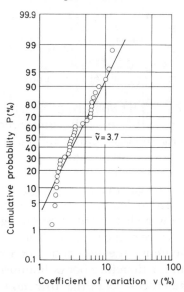

FIG. 18. Distribution of the coefficient of variation v; smooth and hourglass specimens.

It should also be pointed out that each value of S and v in these diagrams was obtained from the P-N data of each test series by a different sample size; this means that each point in the diagrams has a different weight in composing the distribution of S or v, although this difference has not been taken into account in this discussion.

Notched Specimens

A similar analysis was made for the twelve test series of notched specimens. The distribution of σ_{w50}/σ_B was omitted here, however, since the mean fatigue strength σ_{w50} of each kind of notched specimen strongly depends on the stress concentration factor and such a distribution is therefore meaningless. Consequently, only the distributions of the standard deviation S and the coefficient of variation v are shown in Figs. 19 and 20, in the same manner as in Figs. 17 and 18.

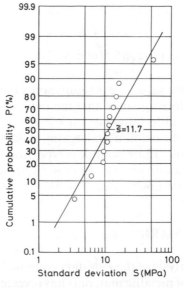

FIG. 19. Distribution of the standard deviation S; notched specimens.

Linear approximations were made for both cases on log-normal probability paper, though the number of points was not large enough for discussion of the exact form of the distributions. The medians are $\tilde{S} \simeq 11\cdot7$ MPa and $\tilde{v} \simeq 5\cdot4\%$, respectively. This value of $\tilde{S}$ is almost equal to the corresponding value for smooth and hourglass type specimens

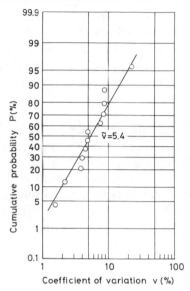

FIG. 20. Distribution of the coefficient of variation v; notched specimens.

given in Fig. 17, but the value of $\bar{v}$ is significantly larger than that shown in Fig. 18. It is therefore apparent that the reduced endurance limit by the notch effect has little influence on the standard deviation of the endurance limit and the median $\bar{S}$ takes almost the same values for both smooth specimens (including hourglass type) and notched specimens; as a result, only the coefficient of variation increases as the endurance limit is lowered by the notch effect.

CONCLUDING REMARKS

The foregoing analysis of the $P-N$ data compiled in the data base on the fatigue strength of metallic materials has revealed a number of basic characteristics of the distributions of fatigue lives and endurance limits of ferrous metals.

The values obtained for the basic parameters characterizing the respective distributions are listed in Tables 7 and 8. In Table 7 the parameter values for complete samples are to be applied to the fatigue life distribution in the finite life range at relatively high stress levels, and those for the censored samples are to be used for the fatigue life

TABLE 7

Weibull parameters for the fatigue life distribution of ferrous metals

Sample	Specimen shape	Shape parameter A	Reduced location parameter $\gamma = C/B$	Relative lower bound of life $C/\tilde{N} \simeq \gamma/(1+\gamma)$	Location parameter C
Complete	SM&RD	1·5	$0·075^a$	$0·07^a$	—
	NT	2·1	$0·12^b$	$0·1^b$	—
Censored	SM&RD	0·85	0·01–0·05	—	10^5
	NT	0·8			

SM, smooth specimen; RD, hourglass type specimen (round bar or plate); NT, notched specimen.
[a]The risk accompanying the value is about $1·8 \times 10^{-3}$.
[b]The risk accompanying the value is about 6×10^{-4}.

TABLE 8

Basic parameters for the endurance limit distribution of ferrous metals

Load	Specimen shape	Mean endurance ratio R_{w50}	Median standard deviation $\tilde{S}$ (MPa)	Median coefficient of variation $\tilde{v}$ (%)
RB, RC, BO	SM, RD	0·50	13·2	3·7
	NT	—	11·7	5·4

RB, uniform rotating bending; RC, cantilever rotating bending; BO, out-of-plane bending of plate; SM, smooth specimen; RD, hourglass type specimen (round bar or plate); NT, notched specimen.

distribution near the endurance limit. In the latter case the value of γ is given by an interval due to some ambiguity as mentioned before. Therefore the value of C is shown in this case as an alternative.

When the distribution function of fatigue life is known for all the stress levels, the distribution function of fatigue strength can be constructed automatically, since the life distribution at a given stress level and the strength distribution at a given life can be transformed into each other [14]. This fact immediately leads to the concept of a failure probability surface covering the entire domain of the stress and fatigue life. A brief explanation of this concept is in order here. The general form of the distribution function of fatigue life is expressed by

$$F_L(N, \sigma) = p(\sigma)F(N, \sigma) + \bar{p}(\sigma)S(N - N_\infty) \tag{6}$$

Here $p(\sigma)$ is the failure probability at a censored number of cycles (10^7 for ferrous metals) and is a function of stress σ; $\bar{p}(\sigma)$ is a complemental probability of $p(\sigma)$. $F(N, \sigma)$ is a distribution function of life at a given stress stress σ and is determined only by the data of failed specimens, assuming them to be a complete sample, as was assumed in the analysis of the censored samples in the previous chapter. $S(N - N_\infty)$ is a step function with a unit step at a sufficiently large number of cycles, N_∞. Therefore the second term in the right-hand side of Eqn. (6) gives the probability of survival at stress σ.

Consequently, the first term in Eqn. (6) provides full information on the distribution within the censored life, and the probability surface for fatigue life and strength within this range can thus be constructed as in Fig. 21, if the central S-N relation is given. The equation of this surface is expressed by

$$F = p(\sigma)F(N, \sigma) \tag{7}$$

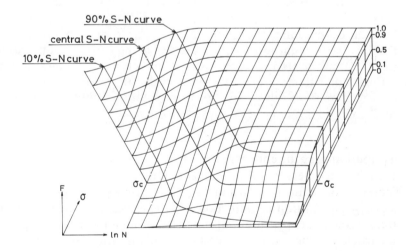

FIG. 21. Illustration of the failure probability surface and P-S-N curves.

Since the central $S-N$ relation is inherent to each material, it should be determined in advance by some other method, either by referring to the $S-N$ data in the data base or by carrying out a conventional fatigue test. In Eqn. (7), the function $F(N, \sigma)$ can be replaced by the Weibull distribution function given by Eqn. (1) for constant σ; $p(\sigma)$ is expressed by normal distribution. Therefore, once the central $S-N$ relation is drawn, the probability surface can be depicted by making use of the values in Tables 7 and 8.

The curved surface schematically shown in Fig. 21 represents the case when the constant value of C is adopted instead of C/B in the range of stress $\sigma < \sigma_c$; in this case, B is also assumed to be a constant equivalent to the value at $\sigma = \sigma_c$. Furthermore, the abrupt change of the shape parameter occurring at $\sigma = \sigma_c$ is smoothed out in Fig. 21. Each curve for a constant σ represents the distribution function of fatigue life, and by our assumption the curves are all parallel in the range where $\sigma > \sigma_c$, while each curve for constant life N provides the distribution function of the strength σ at that life. $P-S-N$ curves with given failure probabilities of 10% and 90% are also shown in Fig. 21. It is a matter of course that the probability surface and the $P-S-N$ curves resting on the surface will change depending on what value of risk probability is taken.

As a final remark, it is noted that the most difficult problem encountered in the foregoing statistical analysis of the fatigue data has been the estimation of the lower limit — or the location parameter — of the life distribution, and attempts are still being carried out by our task force group to find a better and more effective method for this estimation. Efforts are also being made to revise the method of statistical analysis in various aspects, and consequently a more refined recommendation may be proposed in the near future.

ACKNOWLEDGEMENTS

The authors would like to express their appreciation for continuous encouragement and support from repeated discussions with the task force members organized under the auspices of the JSMS Committee on Fatigue. This study was subsidized by the Ministry of Education through a Grant-in-aid for Co-operative Research (A) [60302040], for which grateful acknowledgement is made.

REFERENCES

[1] The Society of Materials Science, Japan, *Data Book on Fatigue Strength of Metallic Materials*, Vols. 1, 2 and 3, JSMS, Japan (1982).
[2] F. A. Bastenaire, *ASTM STP 511*, 3 (1972).
[3] T. Tanaka and T. Fujii, *J. Soc. Mater. Sci., Japan*, 25, 909 (1976).
[4] T. Tanaka and T. Fujii, *Trans. Jap. Soc. Mech. Engrs.*, 42, 2643 (1976).
[5] T. Tanaka, T. Iwaya and T. Sakai, *J. Soc. Mater. Sci., Japan*, 32, 1038 (1983).
[6] T. Sakai and T. Tanaka, *Proc. 24th Japan Congr. Mat. Res.*, 141 (1981).
[7] T. Sakai and T. Tanaka, *J. Soc. Mater. Sci., Japan*, 30, 1211 (1981).
[8] T. Sakai and T. Tanaka, *Proc. Fatigue 84*, II, 1125 (1984).
[9] T. Tanaka and T. Sakai, *Proc. 22nd Japan Congr. Mat. Res.*, 175 (1979).
[10] S. Nishijima *et al.*, *Trans. NRIM*, 19, 43 (1977).
[11] S. Nishijima, *J. Soc. Mater. Sci., Japan*, 25, 53 (1976).
[12] S. Nishijima, *J. Soc. Mater. Sci., Japan*, 29, 24 (1980).
[13] S. Nishijima, *Trans. Jap. Soc. Mech. Engrs.*, 46, 1314 (1980).
[14] T. Tanaka, *Proc. Fatigue 84*, II, 1139 (1984).

Statistical Evaluation of Fatigue Life and Fatigue Strength in Circular-Hole Notched Specimens of a Carbon Eight-Harness-Satin/Epoxy Laminate

TOSHIYUKI SHIMOKAWA and YASUMASA HAMAGUCHI
Second Airframe Division, National Aerospace Laboratory, 7-44-1 Jindaiji-Higashi, Chofu, Tokyo 182, Japan

ABSTRACT

The objective of this study is to make a statistical evaluation of fatigue life and fatigue strength in circular-hole notched specimens of a carbon eight-harness-satin/epoxy laminate. Fatigue damage accumulation and fatigue life distributions were investigated at five test levels of stress amplitude using carefully designed fatigue tests of four-point out-of-plane bending under a room condition of constant temperature and humidity. Sample size was 25 at each stress level. Both the distributional form and the amount of scatter of fatigue life and fatigue strength are mainly discussed based on test results. It is concluded that (1) fatigue strength distributions are practically normal and their standard deviations are constant regardless of fatigue life, and (2) the standard deviation of log-life is approximately in inverse proportion to the slope of the median *S–N* curve on semilogarithmic graph paper.

INTRODUCTION

The application of carbon fiber reinforced plastics (CFRP) to primary structures in aircraft is being actively developed, and a limited number of CFRP primary structures in military aircraft are now in the production stage. Several prototypes of CFRP primary structures in civil aircraft are under flight testing, and these will be in the production stage in the very near future.

CFRP composites have many promising properties as aircraft structural materials. However, on the negative side they have the

159

following weak points: (1) they are weak in compressive, bearing, and impact loads; (2) they weaken by absorbing ambient moisture; (3) they are much less reliable than metals, because they still cannot be produced as homogeneously as metals. Among these weak points, the fatigue reliability of CFRP is one of the most important problems in the design of CFRP primary structures in civil aircraft. Therefore, a statistical evaluation of fatigue life and fatigue strength of CFRP is essential.

A few studies with respect to the amount of scatter and distributional form of CFRP fatigue life have been published [1–10]; however, the total information available is far from sufficient. Most of these data are of small sample size or are limited to a narrow range of stress. The scatter of CFRP fatigue life is generally much larger than that of the fatigue life of aluminium alloys and the CFRP distributional form does not necessarily fit the log-normal or two-parameter Weibull distribution. It is desirable to be able to use a fairly large sample size to investigate the distributions of fatigue life and fatigue strength of CFRP under various combinations of variables, such as laminates, specimen configurations, loading conditions, and environmental conditions.

In a previous study [9, 10] to accumulate such data, the authors fatigue tested sharply notched specimens of a carbon eight-harness-satin/epoxy laminate of three plies under four-point out-of-plane bending, obtained the fatigue life distributions in a wide range of stress, and evaluated both the distributional form and the amount of scatter of fatigue life and fatigue strength. As an extension of the previous study, in the present study the authors used circular-hole notched specimens of a carbon eight-harness-satin/epoxy laminate of six plies and fatigue tested them under experimental conditions similar to those in the previous study. Since many circular holes are bored in aircraft structures, even in those made of CFRP, the data obtained by this kind of specimen configuration are of practical importance. This paper discusses six items from the test results: (1) fatigue damage accumulation and stiffness reduction, (2) fatigue life distributions, (3) the median $S-N$ relation, (4) fatigue strength distributions, (5) $P-S-N$ curves, and (6) comparison of fatigue life data obtained for different panels cured separately. Some of these results are compared with those for sharply notched specimens in the previous study and those of 2024-T4 aluminium alloy [11, 12], which is an aircraft structural metal to be replaced by CFRP.

SPECIMEN AND METHOD OF TESTING

A carbon eight-harness-satin/epoxy laminate (8HS CFRP) was formed using six layers of Mitsubishi Rayon S 410 prepreg sheets with the lay-up sequence of (face/back)$_3$ arranged in the same warp direction. Eight laminate panels were manufactured at Mitsubishi Rayon Ltd. on a hot press separately under an identical cure cycle and trimmed to 340 mm × 280 mm. The design specification and mechanical properties of this laminate are listed in Table 1; the cure cycle is shown in Table 2.

Figure 1 illustrates the specimen configuration. The theoretical stress concentration factor in out-of-plane bending, K_b, is approximately 1·7, which is calculated for a homogeneous isotropic elastic material with Poisson's ratio $v = 0·3$. K_t is used for indicating the specimen configuration. Rectangular specimen blanks were machined by a diamond-wheel cutter, with the warp direction longitudinal. The circular-hole notch was drilled using a HI-ROC drill.

A conventional sub-resonant fatigue machine operating at a frequency of 30 Hz was employed. Specimens were loaded in fully reversed out-of-plane bending by a four-point bending fixture. All fatigue tests with constant amplitude loading were performed with only one fatigue machine and by a single experimenter. This machine had built-in devices to prevent initial overloads. Electricity of constant frequency and voltage supplied from special power equipment having a quartz oscillator was used to drive the synchronous motor of this machine.

The temperature and relative humidity in the laboratory were kept within the ranges 22–24 °C and 50–55%. Other experimental conditions were carefully controlled to obviate additional scatter in fatigue life distribution.

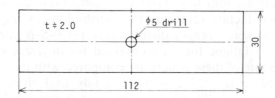

FIG. 1. Specimen configuration (dimensions in mm); $K_b = 1·7$.

TABLE 1

Design specification and mechanical properties of a carbon eight-harness-satin/epoxy laminate of six plies

Prepreg physical properties[a]		Laminate[b] 6 plies in warp direction (face/back)₃			
Prepreg system	S 410 (8-harness-satin)	Property	Mean	c.v.[c] (%)	m[d]
Ply thickness	0·40 mm	Thickness	2·01 mm	1·7	736
Resin content	44 ± 4 wt %	Density	1·59 g/cm^3	0·5	25
Mass/unit	720 g/m^2	Flexural strength[e]	850 MPa	6·3	5
Tensile strength	561 MPa	Flexural modulus[e]	65 GPa	2·8	5
Tensile modulus	59 GPa	Interlaminar shear strength[e]	56 MPa	10·9	5
Flexural strength	941 MPa				
Flexural modulus	63 GPa				
Interlaminar shear strength	76 MPa				

[a] Catalogue data by Mitsubishi Rayon Ltd.
[b] Produced by Mitsubishi Rayon Ltd.
[c] Coefficient of variation.
[d] Sample size.
[e] Tested at Mitsubishi Rayon Ltd.

TABLE 2
Composite cure cycle; cure by a hot press

Process	Pressure (kPa)	Temperature (°C)	Holding time (min)
Cure	—	175	5·5
	686	175	120
Postcure	—	204	120

TEST RESULTS AND DISCUSSION

Some of the test results were compared with those obtained in the previous study [9, 10] using sharply notched specimens of a carbon eight-harness-satin/epoxy laminate (8HS CFRP) of three plies. The stress concentration factor in out-of-plane bending, K_b, of the sharply notched specimens is about 3·8.

Stress Amplitude Representation
Stress amplitude, S, represented by

$$S = 6M/bt^2 \tag{1}$$

for a homogeneous isotropic elastic material, is used in this paper. M is the bending moment loaded to a specimen, b the net specimen width, and t the specimen thickness. The thickness of each specimen was measured and the moment was adjusted so as to give a fixed nominal stress amplitude to the specimen.

Fatigue Damage Accumulation and Stiffness Reduction
The fatigue damage accumulation observed by eye in each specimen of this test series was similar to that observed in the sharply notched specimens in the previous study. Namely, the debonding of a warp and a weft in the surface fabric first occurred at the notch tip. This debonding stopped at the crossover section of both tows, where a warp passed under a weft. This type of debonding propagated gradually warp-by-warp in the weft direction and, macroscopically, the damaged area was formed as a U-shaped area from the root of the circular hole with the bottom of the U-shaped area to the side of the specimen, though a delta-shaped area was formed in the sharply notched specimens. This

behavior was observed on both sides at the notch tips. The growth of delamination was not clearly confirmed. This damage accumulation reduced the stiffness and increased the strain amplitude in the net section. When this strain reached the strain limit of the specimen, the fibers fractured suddenly and simultaneously with the specimen fracture. Gradual propagation of fiber fractures was not found in these tests. Therefore, for these circular-hole notched specimens, fatigue life can be defined by specimen fracture.

Let the stiffness of a rectangular specimen without a notch for out-of-plane bending, namely flexural rigidity, be EI. The residual stiffness ratio, $EI/(EI)_0$, is given [10] as

$$EI/(EI)_0 \doteqdot a_0/a \qquad (2)$$

where a is the amplitude of the reciprocal platen of the fatigue machine. The zero subscript means 'at the start of testing'. The residual stiffness ratio given by Eqn. (2) is used here as the measure of the fatigue damage accumulation in the circular-hole notched specimen.

Figure 2 illustrates the residual stiffness ratio, $EI/(EI)_0$, versus the cycle ratio, n/N, obtained at five test levels of stress amplitude, where n is the number of stress cycles and N the number of cycles to failure. The stiffness of the specimen decreases a few % within 10% of the specimen life ($n/N = 0.1$), followed by a slow, gradual reduction by about 10% until approximately 80% of the specimen life, and then a quick reduction to fracture. The residual stiffness ratio of 80% is judged to be

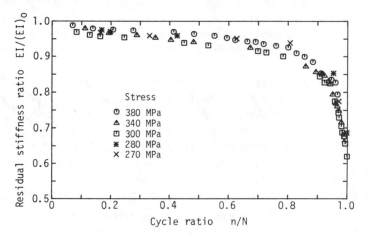

FIG. 2. Stiffness reduction versus cycle ratio.

just before the specimen fracture. The relationship between the residual stiffness ratio and the cycle ratio is independent of the test levels of stress amplitude. The stress-free relationship is favorable in using a specific stiffness reduction ratio as a design criterion.

The stiffness reduction is similar to that of the sharply notched specimens in the previous study; however, the reduction rate of stiffness in Fig. 2 is slower than that of the sharply notched specimens until 80% of the specimen life.

Fatigue Life Distributions

Fatigue life tests were conducted at five levels of stress amplitude. Sample size, denoted by m, was 25 for each stress level. At each stress level, a nearly equal number of specimens were randomly selected from each of five laminate panels, i.e. A, B, C, D and E among a total of eight panels. The test results of fatigue life, N, are shown in Table 3 and in Fig. 3 on log-normal and Weibull probability paper. The plotting positions used in Fig. 3 are the median ranks. It is necessary to use the same plotting positions in comparing the relative goodness-of-fit of the two distribution models. Table 4 lists the median fatigue life and the

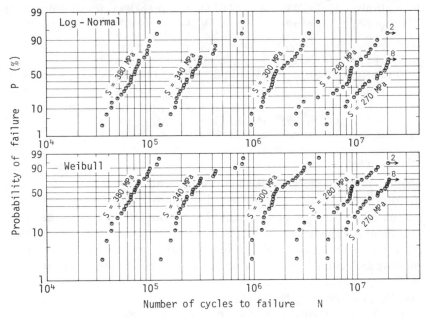

FIG. 3. Fatigue life distributions.

TABLE 3
Fatigue life data

No.	S 380 MPa	P^a	S 340 MPa	P^a	S 300 MPa	P^a	S 280 MPa	P^a	S 270 MPa	P^a
1	34 200	A	125 500	G	954 000	B	2 604 200	A	5 163 100	A
2	37 700	E	156 900	A	959 400	A	2 610 700	B	5 269 900	A
3	42 000	A	173 600	A	1 194 600	E	2 773 400	A	7 863 800	E
4	42 300	A	176 900	B	1 240 500	A	3 093 200	F	8 188 500	F
5	48 200	E	179 400	B	1 250 400	B	4 270 200	C	9 488 600	A
6	52 500	E	188 500	A	1 285 500	E	5 993 800	E	10 171 800	F
7	55 900	D	195 100	B	1 410 500	F	6 460 500	G	10 794 200	E
8	58 300	D	208 100	E	1 495 100	E	7 430 000	B	11 556 200	G
9	61 700	B	211 900	B	1 518 700	A	8 105 000	A	12 161 500	E
10	64 700	D	224 100	A	1 544 700	B	8 497 600	D	13 314 000	B
11	65 000	B	226 000	D	1 551 400	D	8 594 700	B	15 857 200	F
12	65 500	G	253 000	E	1 585 900	B	8 630 400	A	15 996 300	C
13	70 400	A	255 500	F	1 639 100	C	8 816 000	G	17 274 700	B
14	71 000	B	259 000	G	1 683 700	A	8 820 300	E	18 804 000	D
15	72 400	C	274 000	C	1 926 100	D	9 013 400	E	19 514 500	A
16	75 200	C	292 000	E	2 011 300	H	10 124 800	C	19 736 500	B
17	77 400	B	300 400	C	2 171 800	G	10 163 700	F	20 354 500	C
18	77 800	G	302 300	D	2 391 500	C	11 144 000	D	20 504 700^b	G
19	87 800	C	308 300	C	2 569 400	E	12 469 800	E	20 532 300^b	G
20	93 400	H	406 300	E	2 674 900	F	13 232 900	D	20 532 300^b	F
21	94 000	C	420 700	D	2 921 700	C	13 813 400	F	20 536 100^b	D
22	97 200	D	428 500	C	3 046 500	F	15 338 400	D	20 538 200^b	C
23	99 600	F	664 800	G	3 105 500	C	16 040 700	C	20 555 000^b	D
24	116 700	G	776 100	H	3 523 200	D	20 172 300^b	B	20 779 400^b	C
25	122 500	E	793 900	D	4 311 700	D	20 707 100^b	C	20 916 300^b	D

^{a}Parent panel. ^{b}No failure.

estimated parameters for the log-normal and two-parameter Weibull distributions. In the log-normal distribution, the parameters were estimated by the moments method for complete samples and by a graphical plotting technique for incomplete samples, i.e. a combination of the median ranks and the least-squares method. The graphical plotting technique was used in the two-parameter Weibull distribution.

Distributional Form of Fatigue Life

The fatigue life distributions at the four stress levels, $S = 380, 340, 300$ and 270 MPa, fit well with the log-normal distribution. The two-

TABLE 4
Median fatigue life and estimated parameters

Stress, S (MPa)	Sample size, m	Median life, $\tilde{N}$	Log-normal		Weibull	
			Mean, $\bar{x}$	Std. dev., σ_L	Scale, N_c	Shape, α
380	25	70 400	4·830 0	0·147 7	78 900	3·552
340	25	255 500	5·439 8	0·208 6	336 800	2·729
300	25	1 639 100	6·265 1	0·176 8	2 200 400	3·091
280	25(2)[a]	8 816 000	6·922 2	0·277 0	10 628 000	2·081
270	25(8)[a]	17 274 700	7·222 2	0·284 8	19 277 700	2·344

[a] Number of specimens which did not fail.

parameter Weibull distribution is slightly better for the distribution of fatigue life at $S = 280$ MPa. The Kolmogorov–Smirnov (K-S) and chi-square statistics for goodness-of-fit tests in Table 5 provide a comparison of the relative goodness-of-fit of fatigue life data for the log-normal and two-parameter Weibull distributions. The K-S and chi-square statistics verify this tendency. Consequently, for the circular-hole notched specimens, the log-normal distribution is generally superior to the two-parameter Weibull distribution for the distribution model of fatigue life.

TABLE 5
Comparison of goodness-of-fit of fatigue life data for the log-normal and two-parameter Weibull distributions by the Kolmogorov–Smirnov and chi-square statistics

Stress, S (MPa)	Sample size, m	Log-normal		Weibull	
		K-S	$\chi^2(5)^a$	K-S	$\chi^2(5)^a$
380	25	0·088 6	0·80	0·106 8	1·20
340	25	0·166 9	5·20	0·216 0	7·20
300	25	0·147 0	2·80	0·205 8	6·40
280	25(2)[b]	0·160 7	4·00	0·113 8	1·60
270	25(8)[b]	0·069 3	—	0·090 7	—

[a] Number of cells.
[b] Number of specimens which did not fail.

Scatter of Fatigue Life

The amount of fatigue life scatter is commonly measured by the standard deviation of log-life or the shape parameter of the two-parameter Weibull distribution. The estimates of these values, σ_L and α respectively, are listed in Table 4. For the reason described above, fatigue life scatter is discussed here using the standard deviation of log-life. Moreover, the standard deviations of log-life are discussed separately for the two stress regions, i.e. (1) higher than or equal to $S = 300$ MPa and (2) lower than $S = 300$ MPa, because the standard deviations of log-life in Table 4 seem to be different in these two stress regions. The homogeneity of the estimated variances of log-life was tested by Bartlett's chi-square test in the region higher than or equal to $S = 300$ MPa; it was accepted at the 5% significance level. Therefore the amount of fatigue life scatter is considered to be constant in the region higher or equal to $S = 300$ MPa, and the pooled standard deviation of log-life was calculated at 0·18 by the averaging-on-variance basis. Fatigue life data at $S = 280$ and 270 MPa are incomplete, and the estimated standard deviations at these two stress levels are almost equal (approximately 0·28).

Median S–N Curve

The median fatigue lives listed in Table 4 ($K_b \doteqdot 1.7$) are plotted in Fig. 4 on semi-logarithmic graph paper and the median S–N curve is drawn by an eyeball fit. Figure 4 also shows the median S–N curves of

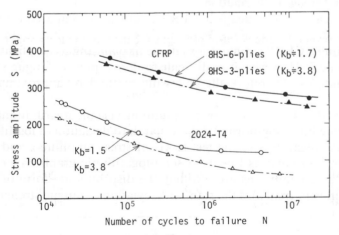

FIG. 4. Comparison of the median S–N curves.

the sharply notched specimens ($K_b \doteq 3 \cdot 8$) in the previous study and those of the similarly notched specimens of 2024-T4 aluminium alloy [12]. The difference in the median fatigue strength between the two specimen configurations of 8HS CFRP laminates is not so large in comparison with that in 2024-T4 aluminium alloy. The median fatigue strengths of the 8HS CFRP specimens are much higher than those of the 2024-T4 aluminium alloy specimens.

An analytical expression of the median S-N relation is necessary both for the estimation of the distribution of fatigue strength and the subsequent discussion of the P-S-N curves. The median S-N curve can be simply approximated by two straight lines on semi-logarithmic graph paper. This analytical method was adopted for the sharply notched specimens in the previous study. The equations of these two straight lines are given in Table 6.

TABLE 6
The median S-N equations on semi-logarithmic graph paper

Stress range (MPa)	Life range (cycles)	Equation
400–300	3×10^4–$1 \cdot 3 \times 10^6$	$S = -58 \cdot 52 \log N + 661 \cdot 3$
300–260	$1 \cdot 3 \times 10^6$–2×10^7	$S = -29 \cdot 03 \log N + 480 \cdot 7$

Fatigue Strength Distributions

The scatter of fatigue life in CFRP is very great compared with that of fatigue strength in general. To use the distribution of fatigue strength in designing a CFRP structure for fatigue or planning its certification test is practical. However, it is difficult to derive precise fatigue strength distributions. Therefore the following method of approximation is used.

Consider the following three conditions: (1) the distributional form of fatigue life is log-normal, (2) the standard deviation of log-life, σ_L, is constant regardless of the test levels of stress amplitude, and (3) the median S-N curve is linear on semi-logarithmic graph paper. When these three conditions are fulfilled, the distributional form of fatigue strength is normal and the standard deviation of fatigue strength, σ_s, is given as

$$\sigma_s = h \sigma_L \qquad (3)$$

where h represented by the absolute value is the slope of the median S–N curve at any fatigue life, N. If the conditions (2) and (3) hold in the vicinity of the test level of stress amplitude, they are considered to be practically sufficient.

As described previously, these three conditions are satisfied sufficiently at the three high stress levels and at the two low stress levels separately. The distribution of fatigue strength is considered to be practically normal, and the standard deviation of fatigue strength, σ_s, can be calculated approximately by Eqn. (3). The values obtained for σ_s are presented in Fig. 5; these do not appear to be a function of fatigue life. The pooled standard deviation of fatigue strength was calculated at 9·6 MPa by the averaging-on-variance basis.

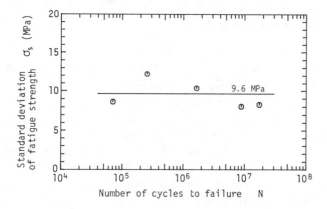

FIG. 5. Estimated standard deviations of fatigue strength.

On the assumption that the distribution of fatigue strength is normal, Fig. 3 indicates that the two-point method [13] (a type of Probit method) is applicable in estimating both the mean value and the standard deviation of fatigue strength directly from the fatigue life data in the life region larger than $2·5 \times 10^6$ cycles. Estimated means and standard deviations of fatigue strength are illustrated in Fig. 6. The estimated means are close to the median S–N curve, and the standard deviation is estimated to be approximately 10 MPa, regardless of fatigue life. In the life region over 8×10^6 cycles in Fig. 5, σ_s is approximately 8 MPa; this agrees with the values in Fig. 6 estimated by the two-point method.

Figures 5 and 6 indicate that $\sigma_s = 10$ MPa is a reasonable estimate for the whole-life region. This is common with the results of the sharply notched specimens used in the previous study and provides an

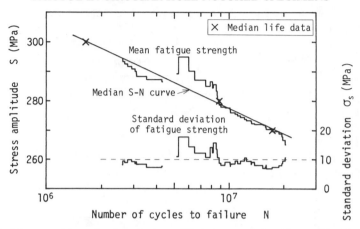

FIG. 6. Mean and standard deviation of fatigue strength estimated by the two-point method.

interesting relationship from Eqn. (3). The standard deviation of log-life, σ_L, is inversely proportional to the slope of the median S-N curve, h. This relationship was previously noted by the authors [11, 12] for 2024-T4 aluminium alloy specimens. Figure 7 presents test results ($K_b \doteq 1\cdot7$) together with the data from the sharply notched specimens ($K_b \doteq 3\cdot8$); a solid line is drawn by introducing $\sigma_s = 10$ MPa into Eqn. (3). Figure 7 shows that Eqn. (3) fits reasonably with the data of σ_L.

The statistical properties obtained for the circular-hole notched specimens, that the distribution of fatigue strength is practically normal and its standard deviation is constant regardless of fatigue life, are similar to those of both the sharply notched specimens and the 2024-T4 aluminium alloy specimens. On the other hand, the standard deviation of fatigue strength of 2024-T4 aluminium alloy was estimated to be approximately 3 MPa, regardless of specimen configuration and fatigue life [12]. Hence, the standard deviation of fatigue strength of 8HS CFRP is approximately three times that of 2024-T4 aluminium alloy. However, Fig. 4 shows that the median S-N curves of 8HS CFRP are sufficiently higher than those of 2024-T4 aluminium alloy. These results would indicate that 8HS CFRP is an excellent substitute material for 2024-T4 aluminium alloy.

P-S-N Curves

The P-S-N curves can be easily derived from the median S-N equation given in Table 6 and from the evaluated distribution of fatigue

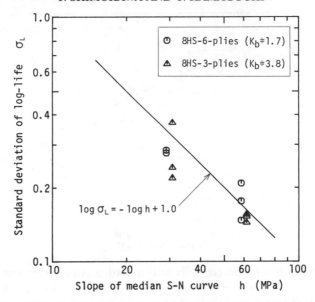

FIG. 7. Relationship between the standard deviation of log-life and the slope of the median S-N curve for two kinds of specimen configurations of 8HS CFRP.

strength, which is normal and has a standard deviation of 10 MPa for the whole-life region. Let the median S-N equation be $S = f(N)$. Then the P-S-N curve corresponding to the probability of failure, P, is represented as

$$S = f(N) + u_p\sigma_s \qquad (4)$$

where u_p is the standard normal variate corresponding to P. Figure 8 shows the derived P-S-N curves and the fatigue life data. These P-S-N curves are a good representation of the fatigue life data and are considered to be sufficient for practical use.

Comparison of Fatigue Life Data Obtained for Five Different Laminate Panels

It is generally recognized that the qualities of CFRP are not as homogeneously controlled as those of metals. Eight laminate panels used in the present paper were produced separately under an identical cure cycle. Three or four specimens for fatigue tests at each level of stress amplitude were sampled from the five panels, A, B, C, D and E, among the total of eight panels. The fatigue life data from these five different

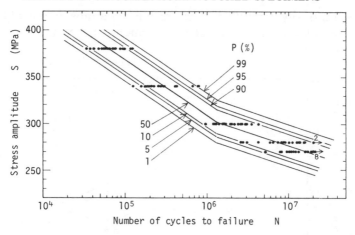

FIG. 8. P–S–N curves and fatigue life data.

panels are compared graphically and statistically at the 5% significance level.

Figure 9 shows the mean S–N relations of these five panels, where the means were calculated on the basis of log-life. Difference in the mean log-lives of the five panels can be recognized intuitively. Statistical tests were carried out on only complete samples of fatigue life data obtained at the three stress levels S = 380, 340 and 300 MPa. At these stress levels four specimens were sampled for each of the five panels. The homogeneity of log-life variances tested by Bartlett's chi-square test was accepted at S = 380 and 300 MPa. At S = 340 MPa it was accepted at the 2·5% significance level. Analysis-of-variance for a single factor experiment indicated that there was a significant difference between the fatigue life data obtained from the five different panels at S = 340 and 300 MPa. At S = 380 MPa it was accepted at the 20% significance level. The correlation between specimen thickness and log-life was also tested statistically. This correlation was not significant at S = 380 or 340 MPa, but positive correlation at S = 300 MPa was not rejected even at the 0·1% significance level.

These results and the fatigue damage accumulation behavior described above lead to the conclusion that a significant difference in fatigue resistance by different panels is caused mainly by the difference in matrix conditions, though the difference in the specimen thickness is by no means negligible. This indicates that the scatter of fatigue life would be improved still more if the quality of the laminate panels was more uniformly controlled. These results are almost identical to those obtained by the sharply notched specimens in the authors' previous study [9, 10].

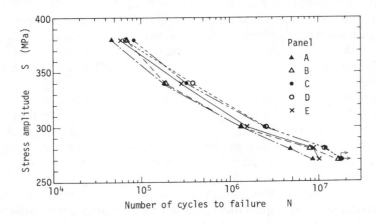

FIG. 9. Mean S-N relations of five different panels on log-life basis.

CONCLUSIONS

(1) According to visual observation of fatigue damage, the propagation of debonding among a warp and a weft in the surface fabric was pronounced, though there was no confirmation of obvious growth of delamination. No fiber fractures were observed during the damage accumulation process even though they were observed at the final specimen fracture.

(2) The stiffness of a specimen decreased a few % within 10% of the specimen life, and showed a slow, gradual reduction by about 10% until approximately 80% of the specimen life, then quickly reduced to fracture. The relationship between the residual stiffness ratio and the cycle ratio was independent of the test levels of stress amplitude.

(3) The log-normal distribution was generally better than the two-parameter Weibull distribution for the distribution model of fatigue life.

(4) In the stress level range $S = 300$–380 MPa, the standard deviation was considered to be constant, roughly 0.18, while at $S = 270$ and 280 MPa it was found to be approximately 0.28.

(5) The distributional form of fatigue strength was considered to be practically normal and the standard deviation of fatigue strength was evaluated to be approximately 10 MPa, regardless of fatigue life. These results were seen in the sharply notched specimens as well.

(6) The results described above provide that the standard deviation of log-life is inversely proportional to the slope of the median $S-N$ curve. This relationship fits the data reasonably in the two specimen configurations of the circular-hole notched and sharply notched specimens.

(7) The $P-S-N$ curves were drawn using the evaluated distribution of fatigue strength, which was normal and had a standard deviation of 10 MPa regardless of fatigue life, and the median $S-N$ relation approximated by two straight lines on semi-logarithmic graph paper. These fit the fatigue life data well and were considered to be sufficient for practical use.

(8) The standard deviation of fatigue strength of 8HS CFRP was approximately three times that of 2024-T4 aluminium alloy. However, the median $S-N$ curves of 8HS CFRP were sufficiently higher than those of 2024-T4 aluminium alloy.

(9) A significant difference in the fatigue resistance of different panels was recognized. The influence of the difference in the specimen thickness, however, was considered to be small.

ACKNOWLEDGEMENTS

The authors wish to express their heartfelt thanks to Dr. Hiroyuki Okamura, Professor of Tokyo University, Dr. Sakae Tanaka, President, and Dr. Masahiro Ichikawa, Professor, of the University of Electro-Communications, for their valuable suggestions and helpful discussion concerning this work.

REFERENCES

[1] J. C. Halpin, *J. Composite Mater.*, **6**, 208 (1972).
[2] H. Nakayasu, Z. Maekawa, T. Fujii and K. Mizukawa, *Proc. 19th Japan Congr. Mater. Res.*, 195 (1976).
[3] J. T. Ryder and E. K. Walker, *AFML-TR-76-241*, WPAFB (1976).
[4] J. N. Yang and D. L. Jones, *J. Composite Mater.*, **12**, 371 (1978).
[5] L. N. Phillips and J. B. Sturgeon, *Int. J. Fatigue*, **1**, 66 (1978).

[6] J. N. Yang, R. K. Miller and C. T. Sun, *J. Composite Mater.*, **14**, 82 (1980).
[7] T. Shimokawa and Y. Hamaguchi, *J. Soc. Mater. Sci., Japan,* **30**, 373 (1981).
[8] W. J. Park and P. Y. Kim, Progress in science and engineering of composites, *Proc. ICCM-IV,* 709 (1982).
[9] T. Shimokawa and Y. Hamaguchi, *J. Composite Mater.*, **17**, 64 (1983).
[10] T. Shimokawa and Y. Hamaguchi, *NAL TR-809T* (1984).
[11] T. Shimokawa and Y. Hamaguchi, *Trans. Japan Soc. Aero. Space Sci.,* **21**, 54, 225 (1979).
[12] T. Shimokawa and Y. Hamaguchi, *J. Eng. Mat. Tech., Trans. ASME,* **107**, 214 (1985).
[13] R. E. Little, *Manual on Statistical Planning and Analysis,* ASTM STP 588 (1975).

Failure Probability Estimation of Impact Fatigue Strength for Structural Carbon Steels and Insulating Structural Materials

NAGATOSHI OKABE*

Heavy Apparatus Engineering Lab., Toshiba Corporation, 1 Toshiba-cho, Fuchu 183, Tokyo, Japan

ABSTRACT

An experimental investigation was conducted on the characteristics of impact fatigue strength, such as stress duration time dependence, the notch effect, influence of heat-treated hardness, and temperature dependence, for structural carbon steels and insulating structural materials, which are usually used in the supporting structural parts or in the operating mechanical components of heavy electrical switchgear. These characteristics were found to be closely related to either the static tensile strength, σ_B, or the heat-treated hardness, H_v, for each material investigated. The dispersions of the impact fatigue strength were also analyzed and found to present approximately a logarithmic normal distribution. The statistics, S_t, which were obtained by a statistical procedure against the ratio, μ, of the experimental values to the estimated values for impact fatigue strength, can be approximately estimated from the statistics S_{σ_B} and S_{H_v} of σ_B and H_v. As a consequence it is proposed here that, considering the estimated S_t as the known value, the impact fatigue strength can be practically evaluated not only by a point estimation but also by an interval estimation based on the failure probability, P_f, the confidence level, $1 - \gamma$, and the sample size, k for either a tensile test or a hardness measurement.

INTRODUCTION

In the design of heavy electrical switchgear such as gas insulating switchgear and vacuum switchgear, supporting structural insulation and operating mechanical parts are formed from structural carbon steels, including heat-treated mechanical parts and such composite materials as reinforced epoxy castings and laminates. These are usually subjected to large impact loads whenever a switching operation is carried out. Therefore the impact fatigue strength to withstand the repetition of these impact loads is significant in the mechanical design.

*Present address: Heavy Apparatus Engineering Lab., Toshiba Corporation, 1-9, Suehio-cho, Tsurumi-ku, Yokohama 230, Japan.

These impact loads are usually single-pulse loadings with a short duration, 20–30 μs to 2–3 ms. Moreover, the repeated interval times of impact loadings are much longer than the duration times. The magnitude of the impact stress often exceeds even the static fracture strength, but fracture is not usually caused during a single cycle or multiple cycles because of the short duration time and the strain rate effect. Generally, mechanical guaranteed lives required for heavy electrical switchgear are less than 10 000–20 000 cycles, but some switchgear for general industry requires the life to be guaranteed to approximately 200 000 cycles. In either case, however, the mechanical strength designs for withstanding the repeated impact loads need fatigue design for a definite life with consideration of scatter of strength and loads.

With this background in mind, the characteristics and dispersion of impact fatigue strength have been investigated experimentally here for the main materials used in the mechanical or insulating structural parts of heavy electrical switchgear. The characteristics obtained by experiment have then been analyzed in relation either to the static tensile strength or to the heat-treated hardness for each material. In addition, both the notch effect and the heat-treated hardness dependence of impact fatigue strength were clarified for structural carbon steels, while the temperature dependence was determined for such composite materials as reinforced epoxy castings and laminates. These serial studies were systematically completed, and it is thus possible to estimate the impact fatigue strength for any sort of mechanical or structural material, from the static tensile strength, σ_B, or the heat-treated hardness, H_v, according to the failure probability, P_f.

This method of evaluation based on the estimation is useful not only for determining the reliability of design against service impact loading of the mechanical and structural parts but also for the application of reliability engineering to quality control and proof tests.

EXPERIMENT

Materials and Specimens

Structural Carbon Steels

The materials used for the experiments described here are four types of commercial structural carbon steels, such as JIS S10C, S20C, S35C and S45C. Smooth and notched specimens were machined to the dimensions and shapes shown in Fig. 1. Either vacuum annealing or normalized heat treatments were performed on the specimens for the experiments on the fatigue notch effect. Quench-tempered heat treatments were performed to obtain specimens of JIS.S45C with four levels of hardness for the experiments on heat-treated hardness dependence. The mechanical properties of these specimens are shown in Table 1.

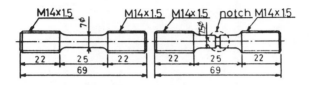

(a) Smoothed specimen   (b) Notched specimen

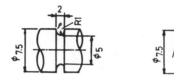

Circumferential U-notch   Circumferential V-notch

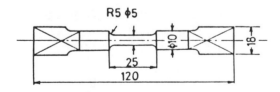

(c) Smoothed specimen

FIG. 1. Shape and dimensions of various specimens of structural carbon steels.

Insulating Structural Materials

The insulating structural materials used in these experiments are of either reinforced epoxy castings or laminates. The epoxy castings are of

TABLE 1
Mechanical properties for various kinds of structural carbon steels

Material	Yielding stress, σ_y(MPa)	Tensile strength, σ_B(MPa)	Elongation, δ (%)	Reduction, ϕ (%)	Vickers hardness, H_v	Heat treatment
S10C	289	381	34·7	74·5	—	Vacuum annealing
S20C	309	438	32·5	63·5	—	Normalizing
S35C	342	533	21·1	54·8	—	Normalizing
S45C	390	634	14·8	46·8	—	Annealing
S45C	374	652	19·9	49·6	166	
S45C	915	982	11·6	44·4	335	
S45C	1 141	1 221	1·9	43·2	417	Quench tempering
S45C	1 496	1 602	0·8	6·9	537	

three kinds: alumina or dry silica or wet silica filler reinforced epoxy resin castings; the latter are of five kinds, either non-woven polyester cloth or glass cloth reinforced epoxy resin laminates. The specimen castings, which have the dimensions and shape shown in Fig. 2(a) and a metal bush with a female screw inserted at each end, were made in a metal mold without machining. In the case of laminates, specimens with the dimensions and shape shown in Fig. 2(b) were cut down from either cylindrical or plate-like laminates. The static tensile strength for each material is shown in Table 2.

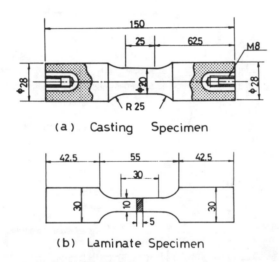

(a) Casting Specimen

(b) Laminate Specimen

FIG. 2. Shape and dimensions of various specimens of insulating structural materials.

Experimental Method

The impact fatigue tests for various structural carbon steels were conducted with three types of impact fatigue testing equipment, as shown in Fig. 3. The type [I] testing device [1, 2] works on the basis of the longitudinal impact phenomenon generated by the combined action of a round bar and a cylinder and can provide a repeated impact tensile stress of rectangular wave-form for each specimen, as shown in Fig. 4(a). This type was used to study both impact fatigue strength and duration time dependence in the ductile fatigue life region. The type [II] testing device [3], which can provide a repeated impact tensile stress as shown in Fig. 4(c) for each specimen by means of a vibro-motor, was

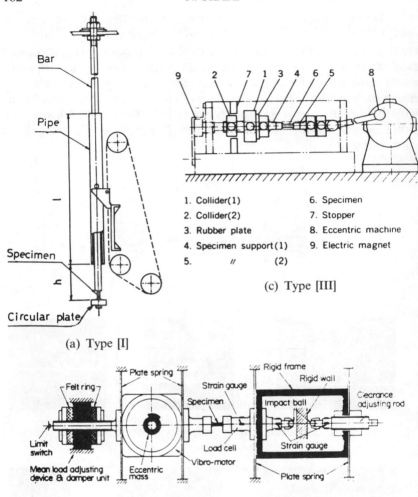

1. Collider(1)   6. Specimen
2. Collider(2)   7. Stopper
3. Rubber plate   8. Eccentric machine
4. Specimen support(1)   9. Electric magnet
5.   "   (2)

(c) Type [III]

(a) Type [I]

(b) Type [II]

FIG. 3. Three types of impact fatigue machines used in the current experiment.

used to study the impact fatigue strength in the brittle fatigue life region. Type [III] [4], which can provide a repeated impact tension–compression stress with any stress ratio from $R = 1$ to $R = 0$, as shown in Fig. 4(b) for the smooth specimen as in Fig. 1(c), was used to study the influence of the stress ratio, R, on the impact fatigue strength in the low cycles life region below several tens of thousands.

TABLE 2
Static tensile strength for various kinds of insulating structural materials

	Forcer	Specimen	Resin[a]	Tensile strength, σ_B (MPa)
Filler-reinforced epoxy resin castings	Alumina filler	[A-I]	Bis-epoxy	83·6
		[A-II]	Bis-epoxy Cyclo-epoxy	68·9
	Silica filler	[S]	Bis-epoxy	85·3
Fiber-reinforced epoxy resin laminates	Non-woven polyester (cylinder)	[P-I]	Bis-epoxy	108·1
		[P-II]	Cyclo-epoxy	80·4
	Glass cloth (cylinder)	[G-I]	Bis-epoxy Cyclo-epoxy	190·0
		[G-II]	Bis-epoxy	217·6
	Glass cloth (plate)	[G-III]	Cyclo-epoxy	314·8

[a] Bis = bisphenol; Cyclo = cycloaliphatic.

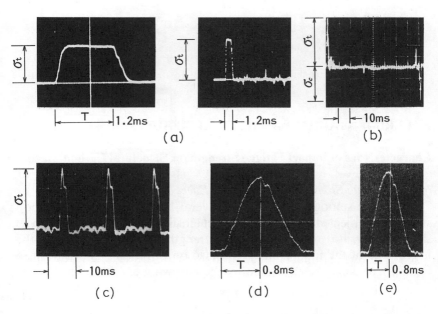

FIG. 4. Examples of measured wave-form of impact stress applied to various specimens.

The impact fatigue tests for insulating structural materials were conducted with a type [I] testing device. The devices for installing a specimen were selected according to whether the specimen is of the casting or laminated type as shown in Fig. 5. Heating to keep a specimen at various testing temperatures was carried out in the high temperature experiments as shown in Fig. 6.

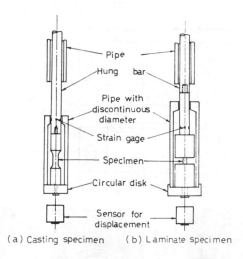

(a) Casting specimen   (b) Laminate specimen

FIG. 5. Devices for installing a specimen of insulating structural materials.

EXPERIMENTAL AND ANALYTICAL RESULTS

Characteristics of Impact Fatigue Strength for Structural Carbon Steels

Impact Fatigue Strength and the Notch Effect

When the smooth specimens of several kinds of structural carbon steel were subjected to repeated impact tensile loads with wave-form as shown in Fig. 4(a), the impact fatigue strength depended not only on the magnitude of the impact stress but also on duration time, as seen in Fig. 7(a), and can be expressed by the following equation [2]:

$$\sigma_t(N_f T)^{m_0} = D_0 \qquad (1)$$

On the other hand, the impact fatigue strength for the notched specimens has similar characteristics, as seen in Fig. 7(b), but can be

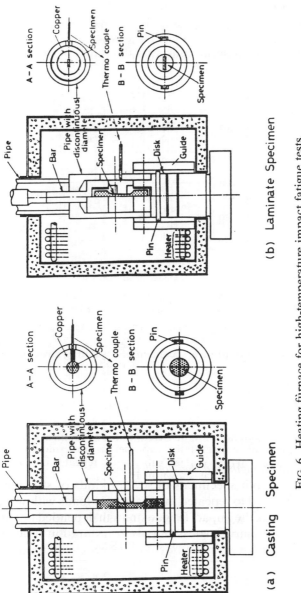

(a) Casting Specimen

(b) Laminate Specimen

FIG. 6. Heating furnace for high-temperature impact fatigue tests.

expressed for any notched specimen with the stress concentration factor, α, as follows:

$$\sigma_t N_f^{m_0} \alpha^{a_0} T^{m_0} = \alpha^{C_0} D_0 \tag{2}$$

where a_0 and C_0 are material constants, the values of which are shown in Table 3.

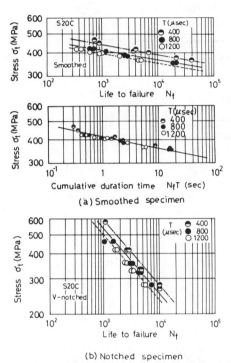

FIG. 7. Diagrams of impact fatigue properties for smooth and notched specimens of structural carbon steels subjected to repeated impact tensile stress as shown in Fig. 4(a).

TABLE 3
Material constants for notch effect on impact fatigue for structural carbon steel JIS.S20C

a_0	C_0
1·26	1·62

Equation (2) becomes Eqn. (1) when $\alpha = 1$, and thus the impact fatigue strength for structural carbon steels can be generally represented for either smooth or notched specimens by Eqn. (2).

Here, m_0 and D_0 are the parameters for impact fatigue strength properties. m_0 is the strength decreasing index and expresses the declination of the S-N diagram. D_0 is the strength constant and determines the strength level.

In the cases with an impact stress wave-form as shown in Fig. 4(b) and (c), the equivalent duration time, $\tilde{T}$, as obtained from the following equation, should be used as the duration time, T, in Eqns. (1) and (2):

$$\tilde{T} = \int_0^t \{\sigma_t/\sigma_{tmax}\}^{1/m_0} \, dt \tag{3}$$

On the other hand, the impact fatigue strength [4] subjected to repeated impact tension–compression loads as shown in Fig. 4(b) can be expressed by using the impact tensile stress, σ_t, and the stress ratio, R ($= \sigma_c/\sigma_t$), on the basis of the extended concept of the endurance limit diagram, as follows:

$$\sigma_t\{(1 - R)N_f^{m_0}\alpha^{a_0}T^{m_0} + RD_1\} = \alpha^{C_0}D_0 \tag{4}$$

where the material constants m_0, D_0 and D_1 have relationships with static tensile properties such as tensile strength, σ_B, elongation, ϕ (%), and ratio of fracture load to maximum load, ρ, as follows:

$$m_0 = 1\cdot087 \times 10^{-3}\,\phi$$

$$D_0 = (0\cdot75 + 0\cdot0025\phi)\,\sigma_B \tag{5}$$

$$D_1 = (D_0/\sigma_B)(1 - 0\cdot01\phi)\,\rho$$

Figure 8 shows the relationships between the experimental values and the values estimated from the static tensile properties (σ_B, ϕ, ρ) for the impact fatigue strength of smooth specimens of structural carbon steels subjected to repeated impact tension–compression loads with a stress ratio of $-1 < R < 0$.

Figure 9 shows the relationships between the experimental values and the estimated values from the static tensile properties (σ_B, ϕ) of each structural carbon steel for impact fatigue strength of each smooth and notched specimen of JIS.S10C, S20C, S35C and S45C subjected to repeated pulsating impact tensile loads with $R = 0$ [5]. As shown in Figs. 8 and 9, impact fatigue strength can be estimated adequately from the static tensile properties by eqns. (4) and (5).

N. OKABE

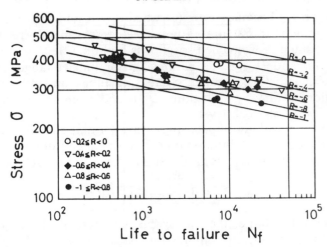

FIG. 8. Diagram of impact fatigue strength for smooth specimens of structural carbon steel JIS.S20C subjected to repeated impact tension–compression loadings with stress ratio of $-1 < R < 0$.

Distribution of Impact Fatigue Strength and the Failure Probability Estimation

The characteristics of impact fatigue strength are shown in Figs. 7–9. The experimental values for impact fatigue strength are dispersed in each S-N diagram, and the ratio, μ_t, of the experimental values to the values estimated from Eqns. (4) and (5) was considered to be the parameter of dispersion in the analysis of the distribution of impact fatigue strength, as follows:

$$\mu_t = \frac{\text{Experimental value}}{\text{Estimated value}} = \frac{\sigma_t\{(1 - R)N_f^{m_0}\alpha^{a_0}T^{m_0} + RD_1\}}{\alpha^{C_0}D_0} \tag{6}$$

On the basis of the order number i, ordered from smaller to larger values, the failure probability, P_f, corresponding to the order number i is given as follows [6]:

$$P_f = (i - 0 \cdot 5)/k \tag{7}$$

where k is the sample size.

Figure 10 shows the relationships between μ_t and P_f on logarithmic normal probability paper. The dispersion of impact fatigue strength for each case shows logarithmic normal distribution.

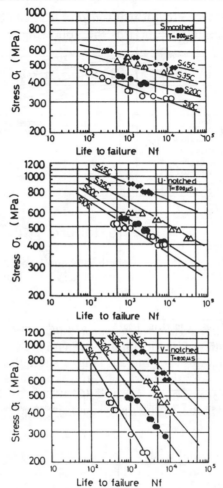

FIG. 9. Diagrams of impact fatigue strength for smooth and various notched specimens of structural carbon steels JIS.S10C, S20C, S35C and S45C.

Therefore, assuming a logarithmic normal distribution, the statistics (μ_t, S_t) of μ_t can be obtained as follows:

$$\widehat{\mu}_t = \exp\left\{\frac{1}{k}\sum \ln \mu_t\right\}$$

$$S_t = \exp\sqrt{\frac{1}{k-1}\sum(\ln \mu_t - \ln \widehat{\mu}_t)^2}$$

(8)

N. OKABE

Then, putting $(\ln \mu_t - \ln \widehat{\mu}_t)/\ln S_t = u, \mu_t = \widehat{\mu}_t S_t^u$ can be obtained, and by substituting μ_t into Eqn. (6), the following equation can be obtained.

$$\sigma_t\{(1 - R)N_f^{m_0 \alpha^{a_0}} T^{m_0} + RD_1\} = \alpha^{C_0} D_0 \widehat{\mu}_t S_t^u \tag{9}$$

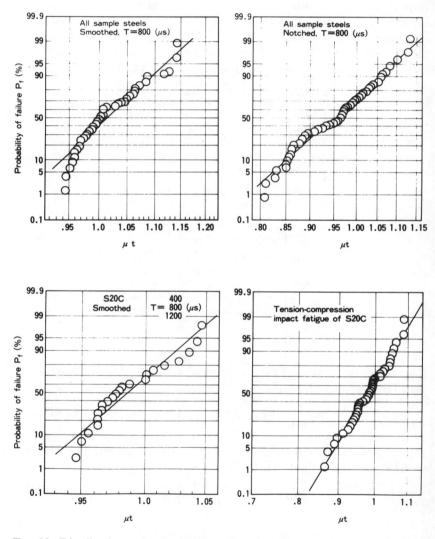

FIG. 10. Distributions of various impact fatigue strengths for structural carbon steel JIS.S20C.

where u is the normal deviation of $\ln \mu_t$ and can be given as the inverse function $\Phi^{-1}(P_f)$ of the normal distribution function $\Phi(u)$ according to the failure probability, P_f [7]. Thus Eqn. (9) can be used to make a point estimation of impact fatigue strength or of impact fatigue lives according to P_f. The S_t representing the magnitude of dispersion for impact fatigue strength naturally depends upon whether the specimen is smooth or notched. However, these values of S_t have a definable relationship with the statistics S_{σ_B}, which represents the dispersion from the median value of σ_B assuming logarithmic normal distribution [5, 7]. This relationship can be expressed by the following equations:

$$S_t = 1\cdot032 S_{\sigma_B} \qquad \text{smooth specimens} \tag{10}$$

$$S_t = 1\cdot062 S_{\sigma_B} \qquad \text{notched specimens}$$

Therefore the S_t can be estimated from S_{σ_B} by considering whether the specimen is smooth or notched.

On the other hand, if the variation coefficient, V_{σ_B}, of σ_B is known, S_{σ_B} can be estimated by the following equation [8]:

$$S_{\sigma_B} = \sqrt{(1 + V_{\sigma_B}^2)} \tag{11}$$

When we try to estimate S_t from V_{σ_B} with the sample size k to obtain σ_B, the following equation should be used to make an interval estimation of impact fatigue strength or of impact fatigue lives according to P_f:

$$\sigma_t \{(1 - R)N_f^{m_0} a^{a_0} T^{m_0} + RD_1\} \alpha^{C_0} D_0 \widehat{\mu}_t S_t^{\Phi^{-1}(P_f)|1 + (\Phi^{-1}(1 - \gamma)/\sqrt{k}|\Phi^{-1}(P_f)|)|} \tag{12}$$

where $1 - \gamma$ is the confidence level.

Characteristics of Impact Fatigue Strength for Structural Carbon Steels with Heat-treated Hardness

Impact Fatigue Strength and Heat-treated Hardness
Under the repeated impact tensile loads shown in Fig. 4(a) and (c), the impact fatigue strength of smooth specimens of structural carbon steels with heat-treated hardness above about H_v 250 can be expressed by the following equation, by considering whether they are in the ductile fatigue life region or the brittle fatigue life region [7, 9], as seen in Fig. 11.

$$\sigma_t (N_f T)^{m_1} = D_1 \qquad \text{ductile life region} \tag{13}$$

$$\sigma_t (N_f T)^{m_2} = D_2 \qquad \text{brittle life region} \tag{14}$$

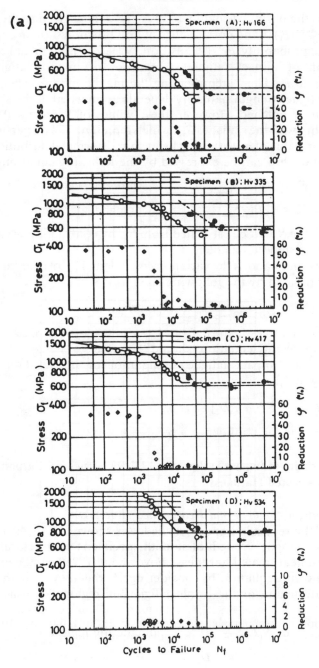

(a)

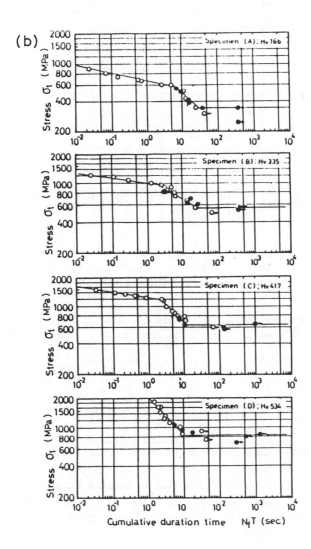

FIG. 11. Diagrams of impact fatigue strength for smooth specimens of structural carbon steel JIS.S45C with four levels of heat-treated hardness. (a) Relation between impact stress σ and cycles to failure N_f. (b) Relation between impact stress σ and cumulative duration time ($N_f T$) until cycles to failure.

The parameters of impact fatigue properties in each life region, (m_1, D_1) and (m_2, D_2), have a relationship with heat-treated hardness, H_v, for each steel as follows [7, 9]:

$$m_1 = m_1^* H_v^{a_1}, \qquad m_2 = m_2^* H_v^{a_2} \tag{15}$$

$$D_1 = D_1^* H_v^{b_1}, \qquad D_2 = D_2^* H_v^{b_2} \tag{16}$$

The endurance fatigue limits, σ_{tw}, and their knee lives, N_w, are also related to the heat-treated hardness, H_v, as follows [7, 9]:

$$\sigma_{tw} = D_3^* H_v^{b_3} \tag{17}$$

$$N_w = \frac{1}{T}\{(D_2^*/D_3^*)H_v^{b_2 - b_3}\}^{1/m_2^* H_v^{a_2}} \tag{18}$$

Here, the material constants $a_1, a_2, b_1, b_2, b_3, m_1, m_2, D_1, D_2$ and D_3 in Eqns. (13)–(18) are shown in Table 4.

TABLE 4
Material constants for influence of heat-treated hardness on impact fatigue for structural carbon steel JIS.S45C with heat-treated hardness above about H_v 250

Ductile fatigue life regions		Brittle fatigue life regions		Endurance fatigue limits	
m_1^*	$3 \cdot 15 \times 10^{-5}$	m_2^*	$1 \cdot 16 \times 10^{-4}$	D_3^*	$3 \cdot 72$
a_1	$1 \cdot 26$	a_2	$1 \cdot 31$		
D_1^*	$1 \cdot 62$	D_2^*	$1 \cdot 37$	b_3	$0 \cdot 876$
b_1	$1 \cdot 11$	b_2	$1 \cdot 16$		

In the cases of impact fatigue for the notched specimens, when the fatigue notch factor is represented by β for the impact fatigue strength, σ_t, at a given life to failure, N_f, and the endurance impact fatigue limit, σ_{tw}, the impact fatigue notch factor, β, can also be expressed by using the stress concentration factor, α, and the heat-treated hardness, H_v, as parameters, as follows:

$$\beta = \alpha^{C_1} H_v^{A(1 - \alpha^{C_2})} \qquad A = A_0 \exp\left\{-\left(\frac{H_v - H_{vL}}{B_0}\right)^{C_3}\right\} \tag{19}$$

where the experimental constants C_1, C_2, C_3, A_0, B_0 and H_{vL} are shown in Table 5.

TABLE 5

Material constants for notch effect on impact fatigue for structural carbon steel JIS.S45C with heat-treated hardness above about H_v 250

C_1	C_2	C_3	A_0	B_0	H_{vL}
2·10	1·65	5·40	0·053 0	332	250

On the other hand, the strength decreasing components, m_{k_2}, for the impact fatigue of notched specimens are almost equal to those for impact fatigue in the brittle fatigue life regions of smooth specimens. Thus the m_{k_2} can also be estimated from heat-treated hardness, H_v, as follows:

$$m_{k_2} = m_2^* H_v^{a_2} \tag{20}$$

In Eqn. (19), β becomes 1 when $\alpha \to 1$. Therefore the impact fatigue strength, σ_t, in the brittle life regions and the endurance impact fatigue limits, σ_{tw}, can be expressed for both smooth and notched specimens as follows:

$$\sigma_t (N_f T)^{m_{k_2}} = D_2/\beta = D_{k_2} \tag{21}$$

$$\sigma_{tw} = D_3^* H_v^{b_3}/\beta \tag{22}$$

For notched specimens with heat-treated hardness above about H_v 500, the life to failure, N_f, against where the impact fatigue strength, σ_t, has stress levels below the tensile fracture strength, σ_f (nominal stress), for each notched specimen, can be expressed as follows:

$$\sigma_t (N_f T)^{m_{k_3}} = D_{k_3} \tag{23}$$

where

$$D_{k_3} = \frac{\sigma_t}{\alpha} \left\{ \frac{\alpha D_2^* H_v^{b_2}}{\beta \sigma_f} \right\}^{(m_1^*/m_2^*) H_v^{a_1 - a_2}} \tag{24}$$

$$m_{k_3} = m_1^* H_v^{a_1} \tag{25}$$

Figure 12 shows the relationships between the experimental and estimated values for the impact fatigue strength of notched specimens with either a U-shaped or a V-shaped circular groove, and suggests that the impact fatigue strength of notched specimens can be approximately estimated from the heat-treated hardness, H_v, by Eqns. (23)–(25). Figure

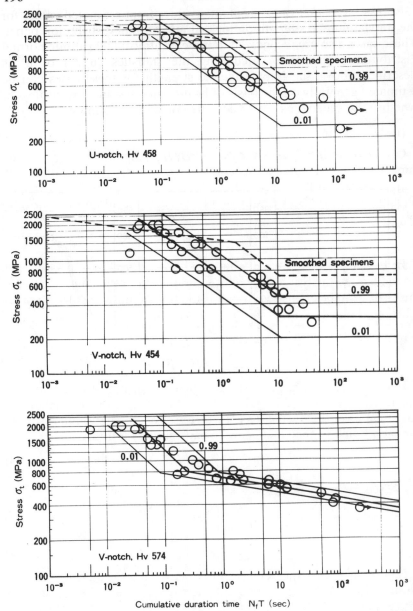

FIG. 12. Diagrams of impact fatigue strength for notched specimens of structural carbon steels with heat-treated hardness.

13 shows the scatter of experimental values from estimated values for the impact fatigue strength on logarithmic normal distribution paper. These cases also are recognized to indicate almost logarithmic normal distribution. Therefore, calculating the statistics (μ_t, S_t) of μ_t as described above, the impact fatigue strength for these cases can be estimated generally on the basis of heat-treated hardness regardless of whether the specimens are smooth or notched, as follows:

(i) Ductile life regions

$$\sigma_t(N_fT)^{m_1^*H_v^{a_1}} = D_1^*H_v^{b_1}\widehat{\mu}_tS_t^u \tag{26}$$

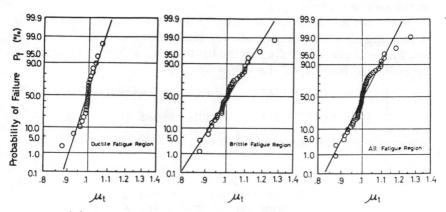

(a) Smoothed specimens of S45C with heat-treated hardness

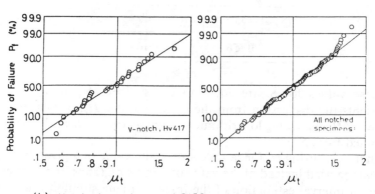

(b) Notched specimens of S45C with heat-treated hardness

FIG. 13. Distributions of various impact fatigue strengths for structural carbon steels with heat-treated hardness.

(ii) Brittle life regions

$$\sigma_t(N_f T)^{m_2^* H_v^{a_2}} = D_2^* H_v^{b_2} \hat{\mu}_t S_t^u \tag{27}$$

for specimens of $H_v \geqslant 500$ and $\sigma_f > \sigma_t > \sigma_{tw}$

$$\sigma_t(N_f T)^{m_1^* H_v^{a_1}} = D_{k_3} \hat{\mu}_t S_t^u \tag{28}$$

(iii) Endurance fatigue limits

$$\sigma_{tw} = (D_3^* H_v^{b_3}/\beta) \hat{\mu}_t S_t^u \tag{29}$$

where u is a probability parameter and is given as $u = \Phi^{-1}(P_f)$ according to the failure probability, P_f.

The statistics, S_t, of μ_t in Eqns. (26)–(29) have a different value depending on whether the ductile life region or the brittle life region is relevant and also depending on the heat-treated hardness, H_v, and can be estimated from H_v by the following equations:

$$S_t = 0.9765 S_0^* H_v^{C_4} \qquad \text{ductile life regions} \tag{30}$$

$$S_t = 1.013 S_0^* H_v^{C_4} \qquad \text{brittle life regions} \tag{31}$$

where the material constants S_0^* and C_4 are indicated in Table 6.

TABLE 6
Material constants for estimating the dispersion of impact fatigue strength from heat-treated hardness

S_0^*	C_4
0·556	0·109

These values for S_t are equal to the statistics S_{H_v} of μ_{H_v} which represents the dispersion from the median value, H_v, of heat-treated hardness, H_v, assuming logarithmic normal distribution, and so can be estimated from S_{H_v}.

When measuring H_v for sample size, k, the value of u given by Eqn. (32) can be used instead of the value of u in Eqns. (26)–(29) in order to make an interval estimation of the impact fatigue strength under the confidence level $1 - \gamma$, according to the failure probability P_f:

$$u = \Phi^{-1}(P_f) \left\{ 1 + \frac{\Phi^{-1}(1 - \gamma)}{\sqrt{k} |\Phi^{-1}(P_f)|} \right\} \tag{32}$$

Characteristics of Impact Fatigue Strength for Insulating Structural Materials

Impact Fatigue Strength at Room Temperature and the Reliability [10, 11]

Figure 14 shows the characteristics of impact fatigue strength for several kinds of insulating structural materials subjected to repeated impact tensile loads as shown in Fig. 4(d) and (e). As observed for filler-reinforced epoxy resin castings in Fig. 14(a), the impact fatigue strength depends on the duration time, T, as well as on the magnitude of the impact tensile stress, σ_t, and can be expressed as follows:

$$\sigma_t N_f^{m_t} T^{n_t} = D_t^* \tag{33}$$

On the other hand, as observed for fiber-reinforced epoxy resin laminates, the impact fatigue strength can be determined using only the magnitude of the impact stress without depending on the duration time, which exists in the range 400–1200 μs, and can be expressed by the following equation:

$$\sigma_t N_f^{m_t} = D_t \tag{34}$$

Now, by using the coefficients of time dependence, ξ_t, to define the dependence of the impact fatigue strength on the duration time, the impact fatigue strength for insulating structural material can be expressed for epoxy castings or epoxy laminates as follows:

$$\sigma_t N_f^{m_t} = \xi_t D_t \tag{35}$$

where ξ_t is given by $\xi_t = (0{\cdot}0008/T)^{n_t}$ and D_t is the strength constant for a duration time of $T = 800$ μs.

The strength constants, D_t, for these cases have a relationship with the static tensile strength σ_B as follows:

$$D_t = a_t \sigma_B^{b_t} \tag{36}$$

where a_t and b_t are material constants and are shown in Table 7. Thus D_t can be estimated easily from σ_B by Eqn. (36).

TABLE 7
Material constants for estimating the strength constants D_t of impact fatigue for various insulating structural materials

a_t	b_t
10·4	1·35

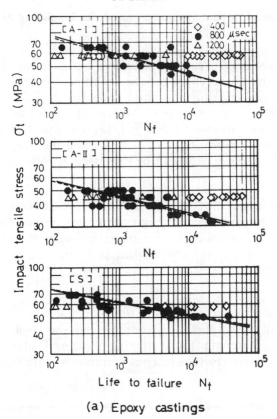

(a) Epoxy castings　　　　　　　　　　　　*contd.*

The strength decreasing components, m_t, differ slightly depending on the kinds of insulating structural material used, but the values of m_t will be in the area of $0 \cdot 11$ in most of the cases.

Figure 15 shows the scatter of experimental values from the estimated values based on the static tensile strength, σ_B, for the impact fatigue strength of each kind of insulating structural material on logarithmic normal distribution paper. These impact fatigue strengths also indicate approximately logarithmic normal distribution. Thus, calculating the statistics $(\widehat{\mu}_t, S_t)$ of μ_t by Eqn. (8) as described above, the point estimation of impact fatigue strength based on σ_B can be made according to the failure probability, P_f, as follows:

$$\sigma_t N_f^{m_t} = \xi_t D_t \widehat{\mu}_t S_t^u = \xi_t D_t \widehat{\mu}_t S_t^{\Phi^{-1}(P_f)} \tag{37}$$

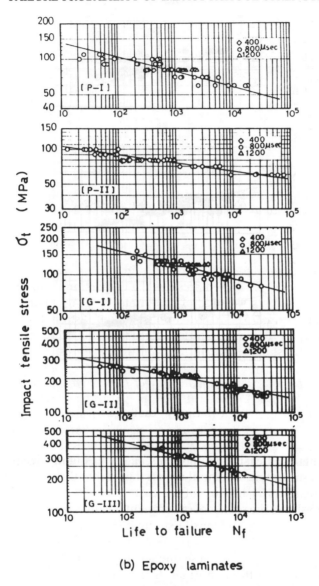

(b) Epoxy laminates

FIG. 14. Diagrams of impact fatigue strength for various insulating materials. (a) Various filler-reinforced epoxy resin castings subjected to repeated impact tensile stress as shown in Fig. 4(d). (b) Various fiber-reinforced epoxy resin laminates subjected to repeated impact tensile stress as shown in Fig. 4(e).

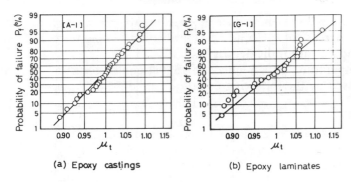

(a) Epoxy castings   (b) Epoxy laminates

FIG. 15. Distributions of impact fatigue strength for insulating structural materials such as alumina-filler reinforced epoxy resin castings and glass-fiber reinforced epoxy resin laminates.

Moreover, these statistics S_t can be estimated as about 1·5 times the statistics, S_{σ_B}, of μ_{σ_B}, which represents the dispersion from the median value, σ_B, for each insulating structural material. Thus, in order to make an interval estimation of the impact fatigue strength under the confidence level $1 - \gamma$ when the sample size for the static tensile tests is k, Eqn. (32) must be used instead of the value of u in Eqn. (37).

Temperature Dependence and Reliability of Impact Fatigue Strength [12]
Figure 16 shows, for several examples, the characteristics of the impact fatigue strength of such insulating structural materials as laminates and several kinds of epoxy castings that were subjected to repeated impact tensile loads at high temperatures. For any material, the impact fatigue strength under a constant high temperature can be expressed as follows:

$$\sigma_t N_f^{m_t} = D_t \tag{38}$$

The dependence of the impact fatigue strength on temperature shows some deviation depending on whether epoxy castings or laminates were used, but it tends in general to decrease with increasing temperature. The temperature dependence of the strength decreasing indexes, m_t, for all insulating materials can be expressed by the following equation (Fig. 17):

$$m_t = m_{\theta1} + m_{\theta2}(T_\theta - 273) \tag{39}$$

where $m_{\theta 1}$ is the strength decreasing index at $0\,°C$, $m_{\theta 2}$ is the increment of m_t per increase of $1\,°C$, and T_θ is the absolute temperature. The values of $m_{\theta 1}$ and $m_{\theta 2}$ are indicated in Table 8.

Temperature dependence of the strength constant, D_t, can be expressed by the following equation (see Fig. 18):

$$D_t = D_\theta^* \exp(-Q_t^*/T_\theta) = D_\theta^* \exp(-Q_t/kT_\theta) \qquad (40)$$

where D_θ^* and Q_t^* are material constants, shown in Table 8, and Q_t with units of active energy is the value which was Q_t^* divided by the Boltzmann constant, k, i.e. $8\cdot617 \times 10^{-5}$ eV.

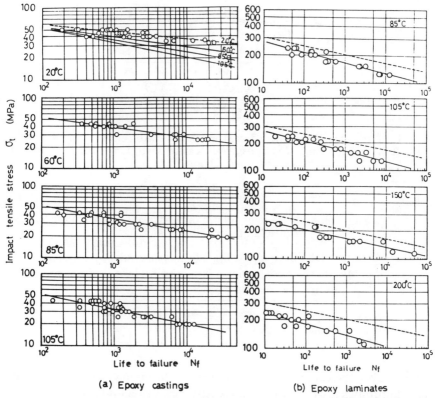

(a) Epoxy castings

(b) Epoxy laminates

FIG. 16. High temperature impact fatigue strength for insulating structural materials. (a) Impact fatigue strength for alumina-filler reinforced epoxy resin castings under elevated temperatures of 60, 85 and 105°C and room temperature. (b) Impact fatigue strength for glass-fiber reinforced epoxy laminates under elevated temperatures of 85, 105, 150 and 200°C.

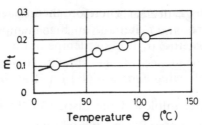

(a) Epoxy castings

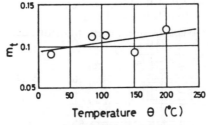

(b) Epoxy laminates

FIG. 17. Dependence of strength decreasing indexes m_t on elevated temperatures. (a) Alumina-filler reinforced epoxy resin castings. (b) Glass-fiber reinforced epoxy resin laminates.

TABLE 8
Material constants for temperature dependence of impact fatigue for insulating structural materials

| Materials | Strength decreasing index | | Strength constants | |
	$m_{\theta 1}$	$m_{\theta 2}$	D_θ^*	Q_t^*
Epoxy castings	0·078 2	$1·16 \times 10^{-3}$	668	424
Epoxy laminates	0·095 0	$1·03 \times 10^{-4}$	217	−162

However, the values of m_t show a distinct difference between castings and laminates for reinforced epoxy samples and tend to increase with increase in temperature in either material. The values of D_t tend to increase for epoxy castings but decrease for epoxy laminates. As described above, the strength constants, D_t, for impact fatigue at room temperature can be estimated from the static tensile strength, σ_B, at room temperature by Eqn. (36). Therefore the impact fatigue strength for insulating structural materials at high temperatures can be estimated practically from σ_B at room temperature as follows:

$$\sigma_t N_f^{m_{\theta 1} + m_{\theta 2}(T_\theta - 273)} = a_t \sigma_B^{b_t} \exp\left\{\frac{Q_t}{k}\left(\frac{1}{293} - \frac{1}{T_\theta}\right)\right\} \tag{41}$$

Figure 19 shows examples of the dispersion of the experimental values from the estimated values based on σ_B for impact fatigue strengths at high temperatures on logarithmic normal probability paper. It is recognized that either case indicates approximately logarithmic normal distribution. Thus, in calculating the statistics $(\hat{\mu}_t, S_t)$ of μ_t, and in investigating the temperature dependence of S_t, the

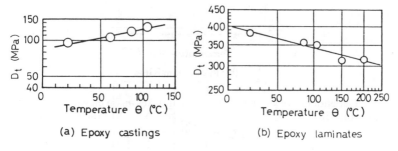

(a) Epoxy castings

(b) Epoxy laminates

FIG. 18. Dependence of strength constants D_t on elevated temperatures.

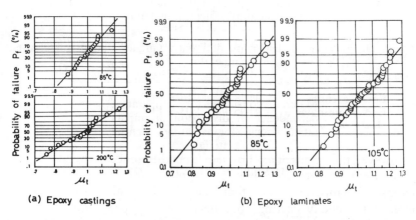

(a) Epoxy castings

(b) Epoxy laminates

FIG. 19. Distributions of high-temperature impact fatigue strength for insulating structural materials.

statistics S_t tend to increase with increasing temperature, as shown in Fig. 20, and can be expressed as follows:

$$S_t = S_\theta \exp(-Q_s^*/T_\theta) = S_\theta \exp(-Q_s/kT_\theta) \qquad (42)$$

where S_θ and Q_s^* are material constants, shown in Table 9, and $Q_s = Q_s^*/k$.

Taking notice of the experimental fact that S_t of the impact fatigue strength can be estimated as approximately 1·05 times S_{σ_B} for static

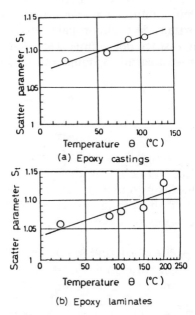

FIG. 20. Dependence of dispersion of impact fatigue strength on elevated temperature.

TABLE 9
Material constants for temperature dependence of dispersion of impact fatigue strength for insulating structural materials

Materials	S_θ	Q_s^*
Epoxy castings	1·26	44·0
Epoxy laminates	1·22	43·7

tensile strength, σ_B, the point estimation of the impact fatigue strength at any high temperature can be made based on S_{σ_B} for static tensile strengths at room temperature according to the failure probability, P_f, as follows:

$$\sigma_t N_f^{m_{\theta 1} + m_{\theta 2}(T_\theta - 273)} = a_t \sigma_B^{b_t} \widehat{\mu}_t (1.05 S_{\sigma_B})^u \exp\left\{\frac{Q_t + u Q_s}{k}\left(\frac{1}{293} - \frac{1}{T_\theta}\right)\right\} \quad (43)$$

where u is a probability parameter, given as $u = \Phi^{-1}(P_f)$. In order to make an interval estimation of the impact fatigue strength under the confidence level $1 - \gamma$, when the sample size for static tensile tests is k, Eqn. (32) must be used instead of values of u in Eqn. (43).

CONCLUSION

Structural carbon steels and insulating structural materials, which are usually used in mechanical parts and insulating structural parts and whose impact fatigue strength to withstand service impact loads is significant for mechanical strength design, were studied here based on the characteristics of impact fatigue strength and on reliability. The results obtained are summarized as follows:

Characteristics of Impact Fatigue Strength for Structural Carbon Steels
  (1) The impact fatigue strength to withstand repeated impact tensile loadings depends not only on the magnitude of the stress but also on the duration time, and can be expressed as follows:

$$\sigma_t N_f^{m_0 a^{a_0}} T^{m_0} = \alpha^{C_0} D_0$$

where α is the stress concentration factor. This equation expresses the impact fatigue strength for smooth specimens, when $\alpha = 1$.
  (2) The impact fatigue strength to withstand repeated impact tension–compression loadings for any stress ratio R can be expressed as follows:

$$\sigma_t\{(1 - R)N_f^{m_0 a^{a_0}} T^{m_0} + R D_1\} = \alpha^{C_0} D_0$$

  (3) The material constants m_0, D_0 and D_1 are closely related to the static tensile properties (σ_B, ϕ, ρ). Thus the impact fatigue strength can be evaluated by an estimation based on these properties.

Characteristics of Impact Fatigue Strength for Structural Carbon Steels with Heat-treated Hardness above about H_v 250

(1) The material constants m_0 and D_0 obviously depend on the heat-treated hardness, H_v. Consequently the impact fatigue strength for smooth specimens can be expressed as follows:

$$\sigma_t (N_f T)^{m_1^* H_v^{a_1}} = D_1^* H_v^{b_1} \qquad \text{ductile life regions}$$

$$\sigma_t (N_f T)^{m_2^* H_v^{a_2}} = D_2^* H_v^{b_2} \qquad \text{brittle life regions}$$

(2) For notched specimens, the impact fatigue strength can be expressed as follows:

$$\sigma_t (N_f T)^{m_2^* H_v^{a_2}} = (D_2^* \beta) H_v^{b_2} = D_{k_2} \qquad H_v < 500$$

$$\sigma_t (N_f T)^{m_1^* H_v^{a_1}} = D_{k_3} \qquad (\sigma_{tw} < \sigma_t < \sigma_f) \quad H_v > 500$$

where α is the stress concentration factor and σ_f is the nominal static fracture strength at the notch bottom; the fatigue notch effect, β, and D_{k_3} are given as follows:

$$\beta = \alpha^{C_1} \exp \left[A_0 (1 - \alpha^{C_2}) \exp \left\{ - \left(\frac{H_v - H_{vL}}{B_0} \right)^{C_3} \right\} \ln H_v \right]$$

$$D_{k_3} = \frac{\sigma_f}{\alpha} \left\{ \frac{\alpha D_2^* H_v^{b_2}}{\beta \sigma_f} \right\}^{(m_1^*/m_2^*) H_v^{a_1 - a_2}}$$

(3) The impact endurance fatigue limits, σ_{tw}, and their knee-point lives, N_w, also depend upon H_v and can be expressed as follows

$$\sigma_{tw} = (D_3^*/\beta) H_v^{b_3}$$

$$N_w = \frac{1}{T} \{ (D_2^*/D_3^*) H_v^{b_2 - b_3} \}^{1/m_2^* H_v^{b_2}} \qquad H_v < 500$$

$$N_w = \frac{1}{T} \{ (D_2^*/D_3^*) H_v^{b_2 - b_3} \}^{1/m_1^* H_v^{b_1}} \qquad H_v > 500$$

Characteristics of Impact Fatigue Strength for Insulating Structural Materials

(1) The impact fatigue strength for reinforced epoxy castings and laminates can be expressed as follows:

$$\sigma_t N_f^{m_t} = \xi_t D_t$$

where ξ_t is the duration time dependence coefficient which is equal to 1 for the casting samples and is given by $\xi_t = (0.0008/T)^{n_t}$ for the laminate samples.

(2) As for the impact fatigue strength for both materials at high temperatures, m_t and D_t depend on temperature and can be expressed as follows:

$$m_t = m_{\theta 1} + m_{\theta 2}(T_\theta - 273)$$

$$D_t = D_\theta^* \exp(-Q_t/kT_\theta)$$

(3) On the other hand, since the relation $D_t = a_t\sigma_B{}^{b_t}$ is operative at room temperature (20°C), the impact fatigue strength at an arbitrary high temperature, $T_\theta(\mathrm{K})$, can be evaluated by an estimation based on σ_B at 20°C:

$$\sigma_t N_f^{m_{\theta 1} + m_{\theta 2}(T_\theta - 273)} = \xi_t a_t \sigma_B{}^{b_t} \exp\left\{\frac{Q_t}{k}\left(\frac{1}{293} - \frac{1}{T_\theta}\right)\right\}$$

Dispersion of Impact Fatigue Strength and its Estimation according to Failure Probability

(1) For both structural carbon steels and insulating structural materials, the impact fatigue strength indicates the dispersion of the logarithmic normal distribution in regard to the estimated values on the basis of either the static tensile properties or the heat-treated hardness for each material.

(2) The statistics, S_t, which were obtained by making a statistical analysis for the ratio, μ_t, of experimental value to estimated value, can be estimated from the statistics (S_{H_v}, S_{σ_B}) of H_v and σ_B respectively. Provided that the estimated S_t is the known value, the impact fatigue strength can be evaluated not only by a point estimation but also by an interval estimation based on the failure probability, P_f, at the confidence level $1 - \gamma$, and the sample size, k, for either tensile tests or hardness measurements.

REFERENCES

[1]   A. Chatani, H. Nakazawa and I. Nakahara, *J. Japan Soc. Mech. Engrs.*, **73**, 1508 (1970).

[2]   T. Uchida, N. Okabe, T. Yano and T. Mori, *J. Soc. Mater. Sci., Japan*, **27**, 1171 (1978).

[3] T. Tanaka and H. Nakayama, *Trans. Japan Soc. Mech. Engrs.*, **29**, 1072 (1973).

[4] N. Okabe, T. Yano and T. Mori, *Proc. 27th Japan Congr. Mater. Res.*, 105 (1984).

[5] N. Okabe, T. Yano and T. Mori, *J. Soc. Mater. Sci., Japan*, **30**, 28 (1981).

[6] T. Shimokawa, NAL., TR464 (1976).

[7] N. Okabe, T. Yano and T. Mori, *4th Int. Conf. Struct. Safety and Reliability*, ICOSSAR '85, Vol. III, 696 (1985).

[8] M. Ichikawa, *Study of Machines*, **35**, 1225 (1983).

[9] N. Okabe, T. Yano, T. Mori, S. Kanayama, H. Nakayama and T. Tanaka, *Fatigue '84*, **II**, 1181 (1984).

[10] N. Okabe, T. Yano, T. Mori and I. Kamata, *J. Soc. Mater. Sci., Japan*, **33**, 41 (1984).

[11] N. Okabe, T. Yano, T. Mori and I. Kamata, *J. Soc. Mater. Sci., Japan*, **31**, 1210 (1982).

[12] N. Okabe, T. Yano, T. Mori and I. Kamata, *J. Soc. Mater. Sci., Japan*, **34**, 333 (1985).

Characteristics of Fatigue and Time-dependent Fracture with Special Reference to the Range of Low Probability Failure

A. TOSHIMITSU YOKOBORI JR and TAKEO YOKOBORI
Department of Mechanical Engineering II, Tohoku University, Aoba, Aramaki, Sendai 980, Japan

ABSTRACT
Attempts have been made (1) to clarify the difference between the underlying concepts of the extreme value theory and that of the stochastic process theory, (2) to show that, nevertheless, distribution functions derived from the stochastic process theory incorporate, as special cases, exactly or nearly the same distribution functions in appearance as derived from the extreme value theory, and (3) to show which of these two theories is valid and useful in what types of failure. For time- or cycle-dependent failure, the stochastic process model may be sufficiently valid. On the other hand, for neither time- nor cycle-dependent brittle fracture, the extreme value theory model may be used as appropriate. Following this, for the prediction of failure probability in the practical, or very low range, the distribution function near the threshold was obtained in terms of the applied stress, σ, the applied repeated cycle, N, and the threshold cycle, N_c. For creep fracture as well, a method has been proposed for estimating the failure probability near the threshold range based on the concept of a two-stage successive stochastic processes model.

INTRODUCTION

The authors have classified problems concerning the reliability of structures into two categories. First, for cases resulting in large risk, we attempted to characterize the range of very small probabilities of failure. Second, for cases in which repair is possible without large risk, a study was carried out to assess the frequency distribution function covering a wide range of probabilities of failure.

This paper concerns the stochastic process theory approach to the reliability of the life of solids and structures subjected to loading. Two kinds of models were used, namely a single stochastic process model and a two-stage model of successive stochastic processes.

Since the 1920s it has been generally accepted [1] that fracture, especially brittle fracture, can be explained from the statistical standpoint, by extreme value theory. On the other hand, early in the 1950s [2-6], the stochastic process theory approach to failure strength and failure life of solids and structures was proposed, and since that time numerous failure problems have been satisfactorily explained by this approach [7-9]; the stochastic process theory was shown to apply not only to the failure of materials, but also to machines and more complex structures, including human beings, in a wide field of reliability engineering [10, 11]. To make matters worse, there appears to be some confusion between the underlying concepts of the extreme value theory and of the stochastic process theory. Therefore attempts have also been made here (1) to make clear the difference between the underlying concepts or theoretical model of the extreme value theory and that of the stochastic process theory for failure, (2) to show that the distribution functions derived from the stochastic process theory include for special cases identical or nearly identical distribution functions in appearance, derived from the extreme value theory, and finally (3) to show which of these two theories is valid and useful in what types of failure.

THE EXTREME VALUE THEORY VERSUS THE STOCHASTIC PROCESS THEORY OF FAILURE

There has not been a clear understanding of the difference between the underlying concepts of the extreme value theory and of the stochastic process theory of failure. Relevant differences and similarities are shown in Table 1.

A presupposition common to both theories is that fracture will occur at the weakest site. However, it should be noted that the two theories and their models are quite different, in both physical and mechanical bases, as follows. In the extreme value theory (i.e. the weakest linkage theory) of failure, the essential model is one where defects with various stress concentrations are randomly distributed and failure is determined by the weakest defect, i.e. the defect with the largest stress concentration at the starting point of load application. In other words, this means that the defect will be the weakest part throughout loading until failure occurs. On the other hand, in the stochastic process theory of failure, the essential model is one where the stress concentration caused by the

TABLE 1
Comparison between the extremal value theory and the stochastic process theory for failure [11]

Approach	Physical or mechanical meaning		Distribution functions obtained
	Similar	Dissimilar	
	Crack initiation, propagation or fracture will occur from the weakest point	Mathematical procedure is different	
Extreme value theory model		The weakest defect at the starting point of loading determines failure, i.e. the defect will be weakest throughout loading until failure occurs; thus the model is only geometrical, and time effects are not in general given physically or mechanically	(1) Weibull distribution function is obtained as the third asymptotic distribution (2) Dual (or doubly) exponential distribution function is obtained as the first asymptotic distribution
Stochastic process theory model		The weakest defect with the largest stress concentration at the starting point of load application does not always remain as the weakest one until failure; thus the model incorporates time effects physically or mechanically	(1) In the cases where transition probability, μ, is expressed as $\mu \simeq Bx^{m-1}$, where x is probability variable, a Weibull distribution function is obtained (2) Log-normal distribution function is obtained approximately in certain ranges for (1) (3) Dual or doubly exponential distribution is obtained approximately for failure strength for the case $\mu \simeq A\exp(\alpha\sigma)$, where σ is applied stress

defects can be changeable throughout the load application period, and therefore the weakest defect that has the largest stress concentration at the starting point of load application does not always remain as the weakest one up to the occurrence of failure. This is why, as is well known, it is difficult when using the extreme value theory of failure to include a time concept physically; that is, it is a static theory in this sense, whereas the stochastic process theory of failure invariably includes a time concept, making it a dynamic theory.

Nevertheless, we will show that distribution functions derived from the stochastic process theory include, as special cases, exactly or nearly the same distribution functions in appearance as derived from the extreme value theory.

STOCHASTIC PROCESS THEORY OF FAILURE AND DISTRIBUTION FUNCTIONS DERIVED FROM THE MODELS: SINGLE STOCHASTIC PROCESS MODEL

Let us assume fracture will occur when a crack initiates in a specimen. In these experiments, the instant of stress application was taken as an origin of time, t. $\mu(t) \, \mathrm{d}t$ is the transition probability, that is the probability of crack initiation within t and $t + \mathrm{d}t$ under the condition that the crack does not initiate until time t. μ is considered as a function of stress where the effects of the environment are negligible. If the process is stochastic, the following equation holds [7–9]:

$$-\mathrm{d}P_{\mathrm{s}}(t)/\mathrm{d}t = \mu(t)P_{\mathrm{s}}(t) \tag{1}$$

where $P_{\mathrm{s}}(t)$ is the non-failure probability until time t. Integrating Eqn. (1) we get

$$P_{\mathrm{s}}(t) \equiv 1 - P_{\mathrm{f}}(t) = \exp\left(-\int_0^t \mu \, \mathrm{d}t\right) \tag{2}$$

where $P_{\mathrm{f}}(t)$ is the probability of failure, i.e. a cumulative distribution function. Since the probability density function, $f(t)$, is related to P_{s} as

$$f(t) = -\mathrm{d}P_{\mathrm{s}}(t)/\mathrm{d}t \tag{3}$$

then from Eqn. (1) we get

$$\mu(t) = -\frac{\mathrm{d} \log P_{\mathrm{s}}(t)}{\mathrm{d}t} \tag{4}$$

The function $\mu(t)$ can be obtained from the tangent to the $\log(1 - P_f)$ versus t curve by Eqn. (4), as shown in Fig. 1 [10, 11].

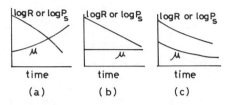

FIG. 1. The change of non-failure probability and transition probability with time; for Weibull distribution corresponding to (a) $m > 1$, (b) $m = 1$ and (c) $m < 1$.

Next, let us note the fundamental concepts, terminology and formulas in reliability engineering compared with a single stochastic process treatment as follows:

$R(t) = 1 - P_f =$ reliability ($= P_s$: non-failure probability)
$f(t) = -dR(t)/dt =$ frequency density function of life
$\lambda(t) =$ mortality or failure rate

$$\lambda(t) = f(t)/R(t) = -\frac{dR(t)}{dt}/R(t)$$

which corresponds to $\mu(t)$ and thus

$$\mu(t) = -\frac{1}{P_s}\frac{dP_s}{dt} = -\frac{dP_s}{dt}/(1 - P_f)$$

The definitions in parentheses are of stochastic process theory.

For comparison, in Fig. 2(a)–(c), the frequency distribution curves, reliabilities and transition probabilities are shown for fracture in unnotched glass under constant stress [2] or in unnotched steels in a corrosive environment under constant stress [12] (Fig. 2(a)), for creep fracture of copper under constant stress [3], or for fatigue fracture of steels under constant amplitude stress [4–6] (Fig. 2(b)), as well as for failure of devices and machines or for death of human beings [13] (Fig. 2(c)). From these curves, it is possible to obtain the reliability, R, or the non-failure probability, $1 - P_f$, and thus the failure rate, λ, or the transition probability, μ. In this way we can see that the transition probability, μ, in general changes with time.

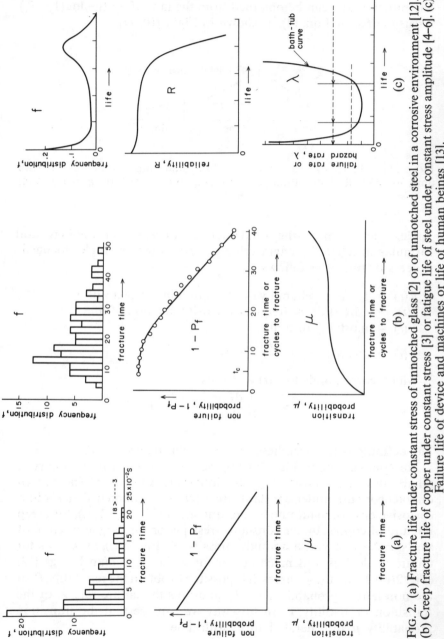

FIG. 2. (a) Fracture life under constant stress of unnotched glass [2] or of unnotched steel in a corrosive environment [12]. (b) Creep fracture life of copper under constant stress [3] or fatigue life of steel under constant stress amplitude [4–6]. (c) Failure life of device and machines or life of human beings [13].

Derivation of the Two-Parameter Weibull Distribution Function

We can see from Eqn. (2) that the Weibull distribution function is obtained [10, 11] as a special case in which μ is denoted as

$$\mu(t) = Bt^{m-1} \tag{5}$$

For this case, from Eqn. (2) we get

$$P_f(t) = 1 - \exp\left(- \int_0^t \mu \, dt\right) = 1 - \exp\left(- \frac{B}{m} t^m\right)$$

that is

$$P_f(t) = 1 - \exp\left[-\left(\frac{t}{t_0}\right)^m\right] \tag{6}$$

where $t_0 \equiv (m/B)^{1/m}$. Note that Eqn. (6) is exactly the same as the Weibull distribution function derived by the extreme value theory as the third asymptotic type distribution function.

In this treatment, the number of load repetitions, N, or strength, $\sigma \, (= \dot{\sigma} t)$, may be used as random variables instead of the time, t, where $\dot{\sigma}$ = loading rate; B is a constant dependent on materials. In most cases of solids and structural failures, the rate of which is controlled by the thermal activation processes (e.g. the yielding of mild steel [14], brittle fracture of steels [15], fatigue crack growth according to the nucleation theory [16, 17], dislocation group dynamics model [18], or the vacancy diffusion model [19]), μ can be given by Eqn. (6) experimentally and/or theoretically.

Derivation of an Approximate Logarithmic Normal Distribution Function

It is shown [20] that a logarithmic normal distribution function, the probability density function of which is given by

$$F(t) \, dt = \frac{\log_{10} e}{\sqrt{2\pi} \, \sigma_*} \frac{1}{(t/t_m)} \exp\left[-\{\log_e(t/t_m)\}^2 / 2\sigma_*^2\right] \, d(\log_e t) \tag{7}$$

where t_m is the median value and σ_*^2 the variance, can be approximated by a Weibull distribution. Let us define the following function:

$$I = 1 + x - e^x \tag{8}$$

If I is expressed in terms of a series expansion for $|x| < 1$, then

$$-\tfrac{1}{2} x^2 = 1 + x - e^x \tag{9}$$

Thus, putting, $x \equiv \log_e(t/t_m)$, and using Eqn. (9), then for the range $1 \gg (1/\sigma_*)\log_e(t/t_m) > 0$, Eqn. (7) can be approximated by the following function:

$$F(t)\,dt = \frac{\log_{10}e}{\sqrt{2\pi}}\,et_m{}^{1/\sigma_*}\left[\left(\frac{1}{t_m}\right)^{1/\sigma_*}\frac{1}{\sigma_*}\left(\frac{t}{t_m}\right)^{1/\sigma_*-1}\exp\left\{-\left(\frac{t}{t_m}\right)^{1/\sigma_*}\right\}\right]d(\log_e t)$$

(10)

The probability density function [ ] in Eqn. (10) is the same type as that obtained by differentiating the Weibull distribution function:

$$P_f = 1 - \exp[-(t/t_m)^{1/\sigma_0}]$$

(11)

where σ_0 is a constant.

In other words, the Weibull distribution function can be approximately expressed in terms of a logarithmic normal distribution function within certain ranges. The experimental data frequently show such a tendency (Figs. 3 and 4). On the other hand, if based on the stochastic

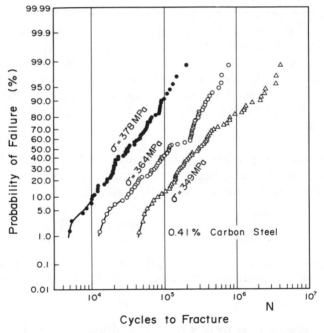

FIG. 3. Fatigue fracture life of 0·41% carbon steel as logarithmic normal plot. (Data: [6])

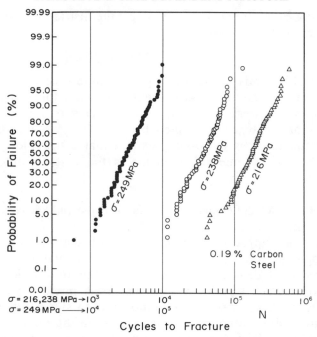

FIG. 4. Fatigue fracture life of 0·19% carbon steel as logarithmic normal plot. (Data: [6])

model, μ can take any function as the case may be, and therefore there might be a case where the logarithmic normal distribution function fits precisely.

DERIVATION OF THE APPROXIMATELY DUAL (OR DOUBLY) EXPONENTIAL DISTRIBUTION FUNCTION

When the rate-determining process in the stochastic process is thermally activated, and the activation energy is a linearly decreasing function of applied stress in terms of the Taylor expansion of series, as is often the case, then the transition probability (or the reaction rate), μ, is given by

$$\mu = A \exp(\upsilon\sigma) \qquad (12)$$

where σ is the applied stress and υ is a constant in relation to activation volume. For fracture under monotonically increasing stress in this case, substituting Eqn. (9) into Eqn. (2), we get the distribution function of fracture strength, σ, as [10, 11]

$$P_f(\sigma) = 1 - \exp B[-B \exp(\upsilon\sigma)] \tag{13}$$

where $B = A/\upsilon\dot{\sigma}$ and $\dot{\sigma}$ is the stress application rate, i.e. $\sigma = \dot{\sigma}t$. That is, Eqn. (13) gives the failure stress distribution function for tensile fracture under constant stress application rate. It is very interesting that equations of the type represented by Eqn. (13) are approximately formally the same as the doubly exponential distribution function derived by the extreme value theory as the first asymptotic type distribution.

The Stochastic Model Coupled with Macroscopic and Microscopic Random Defects

The stochastic process failure model described above need not be a microscopic model. Let us consider a solid containing randomly distributed weak defects. For instance, near the tip of a defect of the macroscopic crack type, there may also be a microscopic random variability, such as stress concentration due to slip bands piled up against the grain boundary. In this case, when the number of weak defects is large, the failure probability, $P_f(\sigma)$, and the mean fracture stress, $\bar{\sigma}_f$, are obtained by using Eqn. (5) [21, 22] as

$$P_f(\sigma) = 1 - \exp\left(-\frac{L_0\sigma^m}{m\dot{\sigma}}\right) \tag{14}$$

and

$$\bar{\sigma}_f = \left(\frac{m\dot{\sigma}}{L_0}\right)^{1/m} \Gamma\left(\frac{m+1}{m}\right) \tag{15}$$

where Γ is the gamma function and

$$L_0 = nn^*A^*R \tag{16}$$

$$R = \int_0^\infty a^{(m+1)/2} \phi(a)\, \mathrm{d}a \tag{17}$$

in which $\phi(a)$ is the probability density function of the half crack length, a; n is the total number of weak macroscopic defects, such as cracks, in the specimen; n^* is the total number of microscopic defects, such as the sites of piling-up of dislocations; A^* is a constant. It can be seen from Eqn. (16) that a size effect is included by the factor nn^* [23].

Table 1 shows the comparison of models and the distribution functions derived both in the extreme value theory and in the stochastic process theory for failure of solids and structures. Table 2 shows the apparent resemblance of the mathematical formulae to each other.

Effect of Stress Application Rate

In most cases there may be a time independent critical stress, beyond which the process will be stochastic or thermally activated. Then Eqn. (15) for the fracture stress, $\bar{\sigma}_f$, becomes

$$\bar{\sigma}_f = \sigma_c + \left(\frac{m\dot{\sigma}}{L_0}\right)^{1/m}\left(\frac{m+1}{m}\right) \tag{18}$$

TABLE 2
Apparent resemblance of the mathematical formula between extreme value theory and stochastic process theory of failure [28]

	Extreme value theory	Stochastic process theory
	Probability density function of crack strength: $f(\sigma_f)$ C.D.F.: $F(\sigma_f) = \displaystyle\int_{-\infty}^{\sigma_f} f(\sigma_f)\,d\sigma$ Weibull (1939) $F(\sigma_f) = 1 - \exp(-\alpha\sigma_f^m)$ $\alpha, m = $ const.	Transition probability: $\mu(\sigma)$ $\mu = \phi(\sigma) = \phi(\dot{\sigma}t)$ Yokobori (1953, 1973) $\mu = L\sigma^\delta$ $L \propto n^*;\ L, \delta = $ const. $n^* = $ total number of microscopic defects
C.D.F. of fracture strength	$Gn(\sigma_f) = 1 - \exp(-n\alpha\sigma_f^m)$ $n = $ total number of macroscopic defects	$D(\sigma_f) = 1 - \exp\left[-\dfrac{L\sigma_f^{\delta+1}}{(\delta+1)\dot{\sigma}}\right]$
Mean value of fracture strength	$\bar{\sigma}_f = \dfrac{1}{(\sigma n)^{1/m}}\Gamma\left(\dfrac{m+1}{m}\right)$	$\bar{\sigma}_f = \left[\dfrac{(\delta+1)\dot{\sigma}}{L}\right]^{1/(\delta+1)}\Gamma\left(\dfrac{\delta+2}{\delta+1}\right)$
		Micro-randomness [23] $L = n^*M,$ $M = $ const. Combined micro- and macro-randomness [21, 22] $L = nn^*A^*R$ $A^*, R = $ const.

$\sigma = $ applied stress, $\sigma_f = $ fracture strength

where σ_c is the critical stress. For $m \gg 1$, the effect of the stress application rate, $\dot{\sigma}$, is not so conspicuous in the actual case. In calculations of failure life under constant stress, the threshold time, t_c, expressed as a deterministic value, may also correspond to this concept.

FATIGUE LIFE CHARACTERISTICS IN THE RANGE OF VERY SMALL PROBABILITY OF FAILURE

Can the extreme value theory be applied physically or mechanically to time-dependent and cycle-dependent failure? An attempt has been made for the case of cycle-dependent failure [24]. For the case of time-dependent failure, however, it is difficult to apply the extreme value theory with physical or mechanical meaning. On the other hand, it has been shown [2, 3, 4, 7–11, 20, 25, 26] that the stochastic process theory can be applied physically and mechanically to time-dependent failure as well as cycle-dependent failure.

There is no formula for failure probability expressed as a function of applied load. Therefore we have attempted here to present the failure probability not only in terms of life but also applied stress. Figures 5 and 6 show Weibull plots for fatigue life data with plain carbon steels obtained experimentally using approximately 100 specimens for each stress level, and taking several stress levels as parameters. Based on the analysis of these data we have obtained the following results.

All series of the data plotted on Weibull paper show the characteristics of having a convex curve upwards (Figs. 5 and 6). That is, the distribution function of fatigue life does not obey a two-parameter Weibull function over the entire scatter range, but has a threshold cycle, N_c (Figs. 5 and 6).

Thus, for the prediction of failure probability, P_f, in the very low range for practical use, a method is proposed here for fitting a three-parameter Weibull function

$$P_f(N) = 1 - \exp\left[-\frac{(N - N_c)^m}{\alpha_0}\right] \tag{19}$$

only for the very low range of P_f. The results are shown in Figs. 7 and 8. α_0, N_c and m are constants. α_0 and N_c were plotted against σ, as shown in Figs. 9 and 10, respectively. From Figs. 9 and 10 we get [20]

$$1/\alpha_0 = \sigma^{25}/A_0{}^m \tag{20}$$

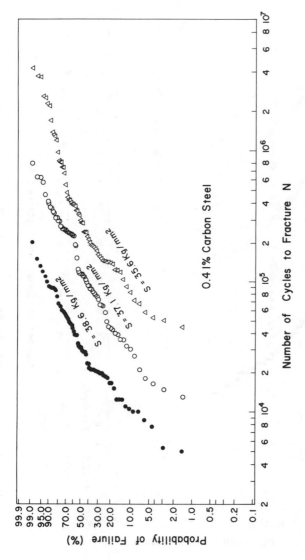

FIG. 5. Fatigue fracture life of 0·41% carbon steel as Weibull plot. (Data: [6])

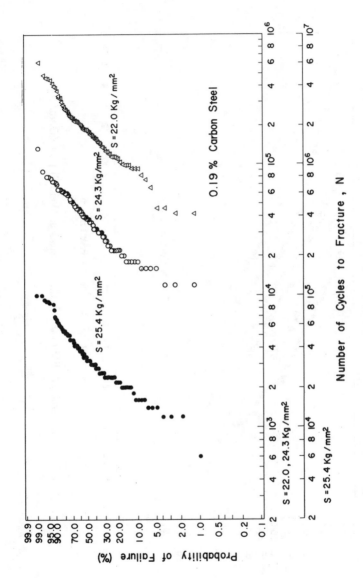

FIG. 6. Fatigue fracture life of 0·19% carbon steel as Weibull plot. (Data: [6])

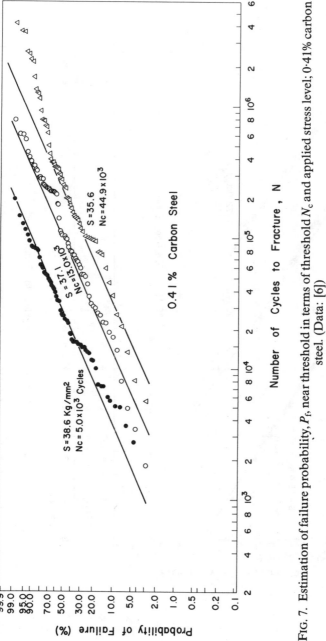

FIG. 7. Estimation of failure probability, P_f, near threshold in terms of threshold N_c and applied stress level; 0·41% carbon steel. (Data: [6])

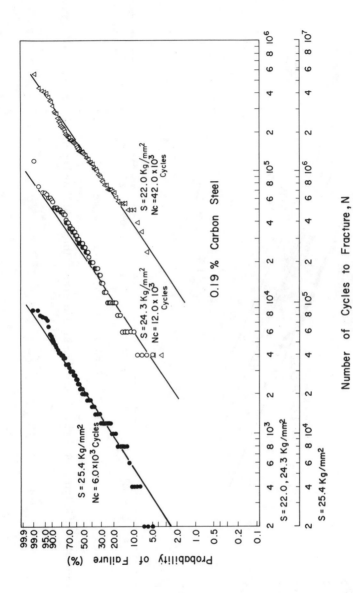

FIG. 8. Estimation of failure probability, P_f, near threshold in terms of threshold N_c and applied stress level; 0·91% carbon steel. (Data: [6])

and

$$N_c = B_1\sigma^{-\gamma} \tag{21}$$

where A_0 and B_1 are constants. The values of m and γ are shown in Table 3. Substituting Eqn. (20) into Eqn. (19), we get the distribution function near the threshold as [10, 11]

$$P_f(\sigma, N) = 1 - \exp\left[-\sigma^{25}\left(\frac{N - N_c}{A_0}\right)^m\right] \tag{22}$$

in terms of the applied stress level, σ, the applied repetition cycle, N, and the threshold cycle, N_c.

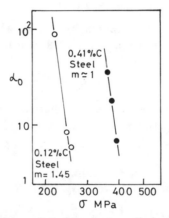

FIG. 9. α_0 plotted against applied stress σ.

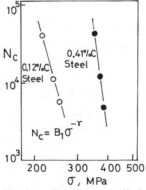

FIG. 10. N_c plotted against applied stress σ.

TABLE 3
Values of m and γ [10]

	m	γ
0·19% carbon steel	1·45	12·5
0·41% carbon steel	1·00	27·4

The meaning of Eqn. (22) may be explained in the following way [10, 11, 20]. The transition probability per unit time, μ_t, is expressed as

$$\mu_t = F(\sigma, t) \tag{23}$$

Let us assume $F(\sigma, t)$ as expressed by $A_0 \sigma^\varepsilon t^\xi$, where A_0, ε and ξ are independent of σ and t. Then μ_N, the transition probability per unit cycle, may be expressed by

$$\mu_N = B_0 \sigma^\varepsilon N^\xi \tag{24}$$

when the triangle stress wave is used, where σ is the stress amplitude. From Eqns. (2) and (24), we get

$$P_f(\sigma, N) = 1 - \exp[-C\sigma^\varepsilon N^\xi] \tag{25}$$

which is essentially the same as Eqn. (22).

For cases with a threshold cycle, N_c, Eqn. (25) reduces to

$$P_f(\sigma, N) = 1 - \exp[-C\sigma^\varepsilon (N - N_c)^\delta] \tag{26}$$

For time-dependent failure, such as creep fracture or corrosion cracking, by similar treatment we obtain

$$P_f(\sigma, t) = 1 - \exp[-C\sigma^\varepsilon t^\delta] \tag{27}$$

or for cases with a threshold time, t_c, Eqn. (27) reduces to

$$P_f(\sigma, t) = 1 - \exp[-C\sigma^\varepsilon (t - t_c)^\delta] \tag{28}$$

For the prediction of failure probability, P_f, in a very small range for practical use, the distribution function near the threshold range is obtained (Figs. 5 and 6) in terms of the threshold cycle, N_c, the applied stress level, σ, and the applied repetition cycle, N, as follows [10, 11]:

$$P_f(\sigma, N) = 1 - \exp\left[-\left(\frac{\sigma}{\sigma_0}\right)^{25}\left(\frac{N - N_c}{N_0}\right)^m\right] \tag{29}$$

where N_0, σ_0 and m are constants [10, 11].

Thus P_f is expressed by Eqn. (29) in terms of both the applied stress, σ, and the repeated cycle, N. In this way, we can construct a P_f-σ-N (or P_f-S-N) diagram as shown schematically in Fig. 11.

Where the time, t, is variable, N, N_c and N_0 are replaced by t, t_c and t_0, respectively, in Eqn. (29) [10].

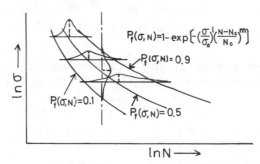

FIG. 11. Schematic illustration of P_f-σ-N (P_f-S-N) diagram obtained from the stochastic process theory of failure [10, 11].

A TWO-STAGE SUCCESSIVE STOCHASTIC PROCESS MODEL

For creep fracture of smooth copper specimens, the life was treated as a two-stage successive stochastic process [5], i.e. the first process starts at the instant of stress application and continues until the initiation of acceleration of creep, the period of which is denoted by t_1; the second process is from the initiation of acceleration of creep to final fracture, the period of which is denoted by t_2 (Fig. 12) [27]. Here, t_1 will be called the damage life and t_2 the propagation life. Regression analysis has shown that there is no correlation between t_1 and t_2 at significance levels of 2·5–5%. Initiation time, t_1, propagation time, t_2, and fracture time, $t_f (= t_1 + t_2)$, plotted on Weibull paper are shown in Figs. 13 and 14 [10]. Similar characteristics are obtained for the other series of data [25]. Based on the analysis of these data we observe the following results.

(1) The distribution functions of damage life, propagation life, and fracture life do not obey a Weibull function over the entire scatter range, but have threshold times t_{1c}, t_{2c} and t_{fc}, respectively (Figs. 13 and 14).

(2) In both damage and propagation processes, the transition probability, μ, increases with time.

(3) The damage life is comparable to the propagation life.

A method is proposed here to estimate the failure probability, P_f, near the threshold range. Since the proposed formula need not be valid over the entire range, the following three-parameter Weibull distribution is proposed near the threshold (Figs. 15 and 16):

$$P_f(t) = 1 - \exp\left[-\left(\frac{t - t_c}{t_0}\right)^{m_0} \right] \tag{30}$$

where t_c is the threshold time and m_0 and t_0 are constants. The method is in good agreement with the data as shown in Figs. 15 and 16.

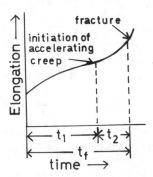

FIG. 12. Schematic illustration of two successive stages in creep fracture of unnotched copper specimen.

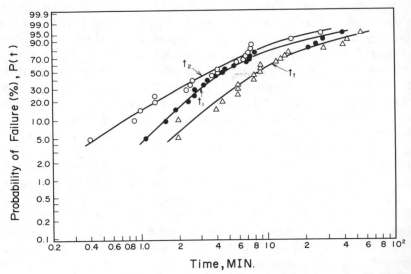

Fig. 13. Creep damage life, its propagation life and creep fracture life of copper under constant stress as a Weibull plot; $\sigma = 240$ MPa, $40\,^{\circ}$C (data from [27]).

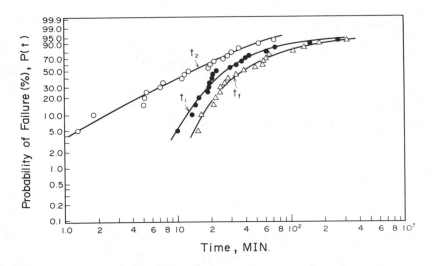

FIG. 14. Creep damage life, its propagation life and creep fracture life of copper under constant stress as a Weibull plot; σ = 250 MPa, 8°C (data from [27]).

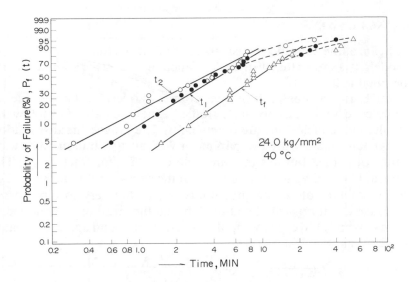

FIG. 15. Estimation of failure probabilities of copper near threshold by three-parameter Weibull distribution; σ = 240 MPa, 40°C. The data are the same as in Fig. 13.

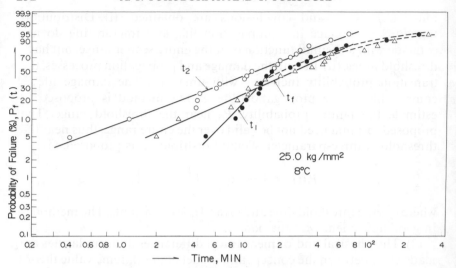

FIG. 16. Estimation of failure probabilities of copper near threshold by three-parameter Weibull distribution; σ = 250 MPa, 8 °C. The data are the same as in Fig. 14.

CONCLUSIONS

From a line of unified considerations of life or strength as stochastic models, it is shown that distribution functions of any type can in general be obtained.

(1) Using a single stochastic process model for fatigue life of unnotched plain carbon steels studied experimentally, the following results and conclusions are obtained. (i) The fatigue life distribution does not coincide with a two-parameter Weibull distribution over a wide range of scatter, but has the threshold cycle, N_c, for fracture. (ii) The threshold cycle, N_c, is given in terms of the stress amplitude, σ. (iii) For the prediction of failure probability, P_f, in the very low range for practical use, the distribution function near the threshold is obtained in terms of threshold cycle, N_c, applied stress level, σ, and applied repeated cycle, N, as

$$P_f(\sigma, N) = 1 - \exp\left[-\left(\frac{\sigma}{\sigma_0}\right)^l \left(\frac{N - N_c}{N_0}\right)^m \right]$$

where σ_0, N_0, l and m are constants.

(2) Using a two-stage successive stochastic process model, the creep fracture life of unnotched OFHC copper is studied experimentally; the

following results and conclusions are obtained. (i) Distribution functions of damage life, propagation life, and fracture life do not coincide with a Weibull function over the entire scatter range, but have threshold times. (ii) In both the damage and propagation processes, the transition probability increases with time. (iii) The damage life is comparable to the propagation life. (iv) A method is proposed to estimate the failure probability, P_f, near the threshold range. The proposed formula need not be valid over the entire range, thus near the threshold, a three-parameter Weibull distribution is proposed, viz.

$$P_t(t) = 1 - \exp\left[-\left(\frac{t - t_c}{t_0} \right)^{m_0} \right]$$

where t_c is the threshold time and m_0 and t_0 are constants. The method is in good agreement with the data.

(3) The physical and/or mechanical differences and similarities are made clear between the concept and model of the extreme value theory and that of the stochastic process theory for failure.

(4) Nevertheless, it is shown that the distribution functions derived from the stochastic process theory include, for special cases, exactly or approximately the same distribution functions in appearance as derived from the extreme value theory, such as Weibull distribution function and doubly exponential function. Moreover, the stochastic process model can include logarithmic normal distribution or any other function in principle.

(5) For time- or cycle-dependent failure, the stochastic process model may be much more valid, and P_f-σ-N or P_f-σ-t diagrams can be constructed based on this model.

(6) On the other hand, the extreme value model may be used for the case when the weakest defect remains exactly as such, i.e. the stress concentration near the defect does not change throughout loading until failure occurs.

(7) The stochastic formula has been derived coupled with macroscopic defects, such as mechanical crack-type, and microscopic defects, such as dislocation piling-up sites, or grain size randomness.

(8) The effect of the stress application rate and the size effect are both included in the proposed approach.

REFERENCES

[1]   F. T. Pierce, *J. Tex. Instit.*, **17**, 355 (1926).
[2]   M. Hirata, *Kikai-no-Kenkyu*, **1**, 231 (1949).

234 A. T. YOKOBORI JR AND T. YOKOBORI

[3] T. Yokobori, *J. Phys. Soc. Japan*, **6** (2), 78 (1951).
[4] T. Yokobori, *Ibia.*, **6** (2), 81 (1951).
[5] T. Yokobori, *J. Phys. Soc. Japan*, **8** (2), 265 (1953).
[6] T. Yokobori, *Rep. Instit. Sci. Technol.*, *Univ. Tokyo*, **8** (1), 5 (1954).
[7] T. Yokobori, *Zairyo Kyodo Gaku*, Giho Do (1955) (in Japanese); *Strength, Fracture and Fatigue of Materials*, (English version) Noordhoff, Groningen, The Netherlands (1965).
[8] T. Yokobori, *Zairyo Kyodo Gaku*, 1st edition, Iwanami Shoten (1964) (in Japanese); *Interdisciplinary Approach to Fracture and Strength of Solids*, (English version) Wolters-Noordhoff, Groningen, Netherlands (1968).
[9] T. Yokobori, *Zairyo Kyodo Gaku*, 2nd edition, Iwanami Shoten (1974) (in Japanese); Russian edition, Naukova, Kiev, (1978).
[10] T. Yokobori and A. T. Yokobori Jr, *Proc. IUTAM Symposium in Weibul Memoriam*, Springer Verlag, 199 (1985).
[11] T. Yokobori and A. T. Yokobori Jr, *4th International Conference on Structural Safety and Reliability. II*, Japan Soc. Mater. Sci., 319–29 (1985).
[12] E. Strecker, D. A. Ryder and T. J. Davies, *J. Iron and Steel Instit.*, **207**, 1639 (1969).
[13] E. Piervshka, *Principles of Reliability*, Prentice-Hall (1963).
[14] A. T. Yokobori Jr, T. Kawasaki and T. Yokobori, *High Velocity Deformation of Solids, IUTAM Symposium*, Springer-Verlag 132 (1979).
[15] T. Yokobori and M. Kitagawa, *Proc. Semi-Internat. Symposium on Exper. Mechanics. II*, Japan Soc. Mech. Engrs., 183 (1967).
[16] T. Yokobori and M. Ichikawa, *Rep. Res. Instit. Str. Fract. Mater.*, Tohoku Univ., Sendai, Japan, **4** (2), 45 (1968).
[17] T. Yokobori, *Physics of Strength and Plasticity*, Argon, Ed., MIT Press, 327 (1969).
[18] T. Yokobori, A. T. Yokobori Jr and A. Kamei, *Internat. J. Fract.*, **11** (5), 781 (1975); **12** (4), 519 (1976).
[19] T. Yokobori and M. Ichikawa, *Rep. Res. Instit. Str. Fract. Mater.*, Tohoku Univ., Sendai, Japan, **6** (2), 75 (1970).
[20] T. Yokobori, A. T. Yokobori Jr and H. Awaji, *J. Japan Soc. Strength of Materials*, **18** (2–3), 43 (1984) (in Japanese).
[21] T. Yokobori and Y. Sawaki, *Internat. J. Fract.*, **9**, 95 (1973).
[22] T. Yokobori, M. Ichikawa and F. Fujita, *Rep. Res. Instit. Str. and Fract. Mater.*, Tohoku Univ., Sendai, Japan, **10** (1), 15 (1974).
[23] T. Yokobori, *J. Phys. Soc. Japan*, **8** (1), 104 (1953).
[24] A. M. Fpreudenthal, *Proc. Roy. Soc. London*, **A187**, 416 (1946).
[25] T. Yokobori, A. T. Yokobori Jr and H. Awaji, *J. Japan Soc. Str. Materials*, **18** (4), 87 (1984) (in Japanese).
[26] T. Yokobori and A. T. Yokobori Jr, *Advances in Fracture Research, Proc. ICF6, Vol. 1*, Pergamon Press, 273 (1985).
[27] T. Yokobori and H. Ohara, *J. Phys. Soc. Japan*, **13** (3), 305 (1958).
[28] T. Yokobori, *Advances in Fracture Research, Vol. 3*, Pergamon Press, 1145–66 (1982).

Statistical Fatigue Life Estimation: The Influence of Temperature and Composition on Low-Cycle Fatigue of Tin-Lead Solders

MAKOTO KITANO, TASUKU SHIMIZU, TETSUO KUMAZAWA
Mechanical Engineering Research Lab., Hitachi, Ltd., 502 Kandatsu, Tsuchiura 300, Japan

and YOSHIYASU ITO
Hitachi Ibaraki Technical College, Hitachi, Ltd., 2-17-2 Nishi Narusawa-cho, Hitachi 316, Japan

ABSTRACT

The reliability of semiconductor devices depends upon the resistance of the solder joints to thermal fatigue failure. Tin–lead solders are the most generally used materials in the fabrication of semiconductor devices. In order to develop a life prediction technique for solder joint fatigue failure, torsion fatigue tests were conducted on hollow cylindrical solder specimens for a series of samples of different compositions. The results are summarized as follows. (1) The low-cycle fatigue life of any solder decreases with an increase of temperature. However, the influence of temperature on fatigue life differs with composition of the solder material. (2) The optimum composition of solder for low-cycle fatigue differs with the magnitude of the strain range. For this reason, the choice of composition should be made in accordance with the strain range the solder will experience. (3) A lack of defects in the solder joint interface has no positive influence on the fatigue strength of solder. (4) The fatigue life of solder joints with defects can be obtained by using the defect area rate. Thus, the probability distribution of fatigue life can be estimated from the distribution of the defect area rate.

INTRODUCTION

Pb–Sn solders are the most generally used materials in the fabrication of semiconductor devices because of their pliability, low melting point, and ease of fabrication. Pb–Sn solders do not damage such delicate semiconductor device parts as silicon chips because they

235

remain pliable at room temperature and have low melting points. Also, several parts can be soldered simultaneously by using re-flow soldering techniques.

Semiconductor devices consist of several parts with different coefficients of linear expansion that must be soldered to each other. Figure 1 shows the schematic structures of two semiconductor devices. When these solder joint assemblies are subjected to cyclic temperatures, expansion mismatch produces cyclic shear strain in the solder layers. When these temperature cycles are large, thermal fatigue is caused in the solder layers. The heat generated in Si chips during operation passes through the solder layers and dissipates to the substrate. The temperature difference between the Si chip and the substrate per unit power, i.e. the thermal resistance, increases with thermal fatigue failure caused in the solder layers. Figure 2 shows the increase in thermal

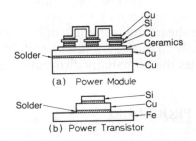

FIG. 1. Schematic structure of semiconductor devices.

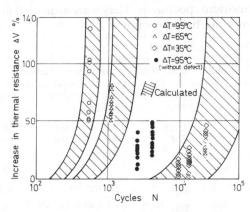

FIG. 2. Increase in thermal resistance in the packages due to thermal fatigue failure.

resistance in the packages due to thermal fatigue failure caused by on/off cycles [1].

The long-term reliability of semiconductor devices is of prime interest to manufacturers and consumers, stimulating studies of solder thermal fatigue in recent years. In particular, many studies have been concerned with the types of solder interconnections used between Si chips and substrates in computers and requiring extreme reliability [2–5]. Coombs [6] compared the results of low-cycle fatigue tests conducted in bulk specimens and lap-type solder joints. Rathore *et al.* [7] conducted bending fatigue tests for several Pb-rich solders and studied the influence of testing frequency and temperature on fatigue lives.

Studies concerned with the low-cycle fatigue of Pb–Sn solders [2–7], however, are inadequate. The influence of temperature, composition, and joint interface has not been well understood. Therefore we conducted torsion fatigue tests on several series of solders to determine the influence on the shear fatigue strength of Pb–Sn solders. This paper outlines the experimental procedure, discusses the dispersion of the test results, and estimates the probability distribution of fatigue life.

EXPERIMENTAL PROCEDURE

The chemical compositions of the solders tested are shown in Table 1. These solders were cast into 30 mm diameter cylinders and machined into hollow cylindrical specimens. They were annealed at 100 °C for 1 h. Figure 3(a) shows the bulk specimen.

TABLE 1
Chemical composition

Sample	Sn (%)	Sb (%)	Cu (%)	Bi (%)	Pb (%)
Pb	0·28	0·12	0·01	0·01	bal.
95Pb–5Sn	5·4	<0·01	<0·01	<0·01	bal.
70Pb–30Sn	32·5	0·01	<0·01	<0·01	bal.
60Pb–40SN	41·5	0·01	<0·01	<0·01	bal.
40Pb–60Sn	61·0	0·01	0·02	<0·01	bal.
20Pb–80Sn	bal.	0·01	0·02	<0·01	20·0
Sn	bal.	0·01	0·02	0·01	0·01
Solder joint specimen	bal.	0·01	0·02	<0·01	42·0

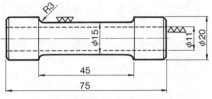

(a) Bulk Specimen

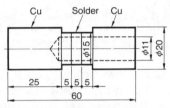

(b) Solder Joint Specimen

FIG. 3. Test specimens; all dimensions in mm.

The solder joint specimens are produced as follows. Copper (OFHC) specimens are heated to 200°C and covered at the solder joint portion by a stainless steel jig. Following this, 40Pb–60Sn solder heated to 220°C is cast into the copper specimens and bored to leave ring-shaped solder joints 5 mm in depth.

Torsion fatigue tests were carried out with a servo-hydraulic fatigue testing machine under controlled torsion angles. The test frequency was 1·0 Hz; the test wave-form was a symmetrical triangular wave. The number of cycles to failure, N_f, was defined by a 20% decrease in the shear stress range from the values at $\frac{1}{2}N_f$.

BULK SPECIMEN RESULT

The universal slope method [8] was applied to the relationship between the shear strain range, $\Delta\gamma$, and the number of cycles to failure, N_f:

$$\Delta\gamma = \Delta\gamma_e + \Delta\gamma_p$$
$$= C_e \cdot N_f^{-k_e} + C_p \cdot N_f^{-k_p} \tag{1}$$

where $\Delta\gamma_e$ is the elastic strain range, $\Delta\gamma_p$ is the plastic strain range and C_e, C_p, k_e and k_p are material constants. Figure 4 shows the result for

95Pb–5Sn solder at room temperature. Equation (1) approximates the test results shown in Fig. 4. The results for other compositions and temperatures are similar to the results shown in Fig. 4, except those for 95Pb–5Sn solder at 150 °C.

Table 2 shows the values of C_e, C_p, k_e and k_p at room temperature. The value of k_p is minimum for eutectic solder (40Pb–60Sn). An attempt was made to apply equations proposed by other researchers to the relationships between C_e, C_p, k_e and k_p and to other mechanical properties for the normal stress range. However, none of these proved to be appropriate.

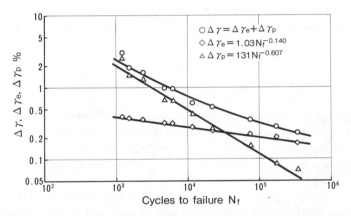

FIG. 4. Application of the universal slope method to the fatigue test result of 95Pb–5Sn solder at room temperature.

TABLE 2
Low-cycle fatigue properties of Pb–Sn solders at room temperature

Sample	C_e(%)	C_p(%)	k_e	k_p
Pb	1·73	1133	0·229	0·852
95Pb–5Sn	1·03	131	0·140	0·607
70Pb–30Sn	1·14	85·7	0·110	0·550
60Pb–40Sn	1·07	66·1	0·113	0·543
40Pb–60Sn	0·427	23·9	0·065	0·411
20Pb–80Sn	1·04	423	0·144	0·874
Sn	0·555	141	0·172	0·706

$$\Delta\gamma = C_e \cdot N_f^{-k_e} + C_p \cdot N_f^{-k_p}$$

INFLUENCE OF TEMPERATURE

The fatigue strengths of 95Pb–5Sn solder and of 40Pb–60Sn solder from −60°C to 150°C are shown in Fig. 5 and Fig. 6, respectively. The fatigue lives of both solders decrease with an increase in temperature, but the influence of temperature is greater for 95Pb–5Sn solder, which has the higher melting point. The fatigue life of 95Pb–5Sn solder at 150°C decreases significantly at the lower strain range.

Figure 7 shows the fatigue cracks of 95Pb–5Sn solder both at room temperature and at 150°C. This figure shows the growth of a transcrystalline crack at room temperature and that of an intercrystalline crack at 150°C. Growth of the transcrystalline crack was also observed at 90°C. Figures 8 and 9 show the stress range change of 95Pb–5Sn

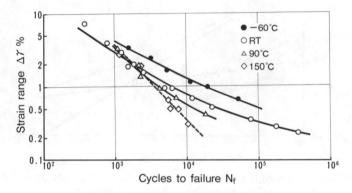

FIG. 5. Fatigue strength of 95Pb–5Sn solder.

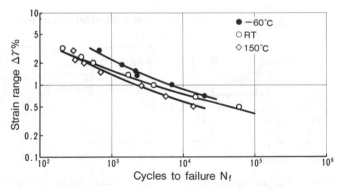

FIG. 6. Fatigue strength of 40Pb–60Sn solder.

solder during fatigue testing at room temperature and at 150°C, respectively. The stress range shows only slight changes at room temperature before failure; however, 95Pb-5Sn solder is softened at 150°C under cyclic strain. The 40Pb-60Sn solder exhibits behavior similar to the 95Pb-5Sn solder, but is not softened as much as 95Pb-5Sn solder at 150°C.

The fatigue behavior of 95Pb-5Sn solder at 150°C is different from those of the other compositions and temperatures described above. This

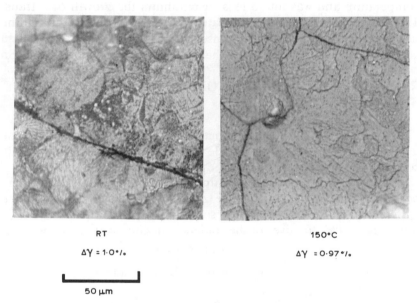

RT

ΔY = 1·0°/₀

150°C

ΔY = 0·97°/₀

50 μm

FIG. 7. Initial fatigue cracks in 95Pb-5Sn solder.

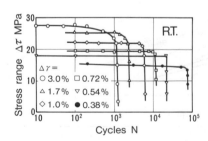

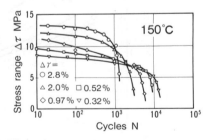

FIG. 8. Stress range change of 95Pb-5Sn solder at room temperature.

FIG. 9. Stress range change of 95Pb-5Sn solder at 150°C.

behavior may be caused by the phase transformation of the solder [7]. The microstructure of 95Pb–5Sn solder at room temperature is eutectic and becomes α-single-phase above 97°C, as shown in the phase diagram of Pb–Sn alloy (Fig. 11). For this reason, the slope for 95Pb–5Sn solder in the $\Delta\gamma$–N_f diagram at 150°C is close to the slope for pure Pb (Figs. 5 and 10).

The influence of temperature was studied for other compositions. It was found that, with an increase of temperature, the fatigue lives of any solders decrease. However, the influence of temperature differs with composition and was found to be a minimum for 60Pb–40Sn solder. The temperature influence decreases with an increase in strain range.

RELATIONSHIPS BETWEEN COMPOSITION AND FATIGUE STRENGTH

Figure 10 shows the fatigue strengths of Pb–Sn solders tested at room temperature. The slope is gentle for eutectic solder. The result corresponds to the minimum value of k_p, as shown in Table 2.

Figure 11 shows the relationship between fatigue lives and composition. In the higher strain range ($\Delta\gamma \geqslant 2\%$), the life of eutectic solder is shortest and the life of Pb-rich solder is longer. In the lower strain range ($\Delta\gamma \leqslant 1\%$), the lives of the solders with compositions between A

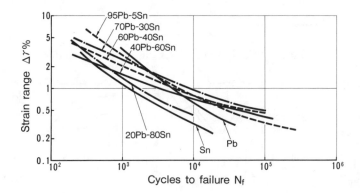

FIG. 10. Fatigue strength of Pb–Sn solders at room temperature.

and *B* in the phase diagram are relatively long. The relationship between life and composition is convex. Figure 12 shows the relationship between composition and shear stress range at $\Delta\gamma$ = 2%. The relationship between composition and fatigue lives in the lower strain range exhibits the same trend as the relationship between composition and shear yield stress.

On the basis of these test results, it would appear advantageous to use eutectic solder when it is necessary to withstand high-cycle fatigue, and Pb-rich solders for low-cycle fatigue.

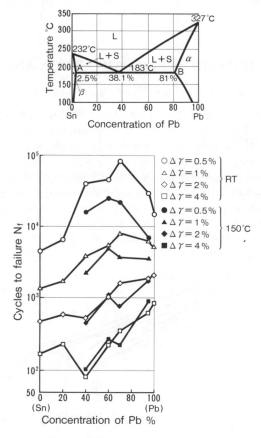

FIG. 11. Relationship between composition and fatigue strength.

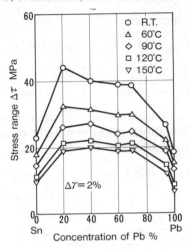

FIG. 12. Relationship between composition and stress range at $\Delta\gamma = 2\%$.

SOLDER JOINT SPECIMEN RESULTS

Figure 13 shows the results of fatigue tests on solder joint specimens. The results for bulk specimens of the same composition are also plotted in this figure.

By application of Eqn. (1) to these results, the following estimated values for the material constants $\widehat{C}_e, \widehat{C}_p, \widehat{k}_e$ and $\widehat{k}_p$ were obtained by using a least-squares method.

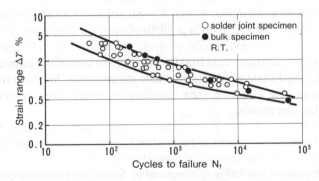

FIG. 13. Fatigue strength of solder joint specimen.

For solder joint specimens:

$$\widehat{C}_e = 0.817\%, \qquad \widehat{C}_p = 15.38\%, \qquad \widehat{k}_e = 0.0742, \qquad \widehat{k}_p = 0.415 \qquad (2)$$

For bulk specimens:

$$\widehat{C}_e = 0.427\%, \qquad \widehat{C}_p = 23.9\%, \qquad \widehat{k}_e = 0.0648, \qquad \widehat{k}_p = 0.411 \qquad (3)$$

The residual sums of the squares of the experimental values, S_e and S_p, and the standard deviations of the residuals, σ_e and σ_p, are calculated from the following equations:

$$S_e = \Sigma(\ln\gamma_e - \ln\widehat{\gamma}_e)^2, \qquad S_p = \Sigma(\ln\gamma_p - \ln\widehat{\gamma}_p)^2 \qquad (4)$$

$$\widehat{\gamma}_e = \widehat{C}_e \cdot N_f^{-\widehat{k}_e}, \qquad \widehat{\gamma}_p = \widehat{C}_p \cdot N_f^{-\widehat{k}_p} \qquad (5)$$

$$\sigma_e = \sqrt{\frac{S_e}{n-1-p}}, \qquad \sigma_p = \sqrt{\frac{S_p}{n-1-p}} \qquad (6)$$

where n is the number of samples and p is the number of predictor variables ($=1$). The calculated results are as follows.

For solder joint specimens:

$$\sigma_e = 5.48 \times 10^{-2}, \qquad \sigma_p = 0.1191 \qquad (7)$$

For bulk specimens:

$$\sigma_e = 4.89 \times 10^{-3}, \qquad \sigma_p = 3.66 \times 10^{-2} \qquad (8)$$

From these equations, it is clear that dispersion of the experimental values of solder joint specimens is from three to ten times larger than those for the bulk specimens.

The causes for this dispersion are considered to be as follows.

(1) *Dispersion by defect.* When Pb–Sn solder is joined to copper, the melting solder wets the surface of the copper sufficiently to allow the formation of intermetallic compounds such as Cu_3Sn or Cu_6Sn_5. If there is any oxide coating or residual flux on the surface of the copper, defects will be formed instead of intermetallic compounds. With an increase in defect area, the effective area resisting external force decreases, and fatigue life decreases.

(2) *Dispersion by voids in solder.* When solder solidifies, voids are formed from air dissolved in the melting solder. Accordingly, fatigue life decreases in the same way as described in (1).

(3) *Dispersion by change of composition.* Since intermetallic compounds are formed with Cu and Sn, but without Pb, it is considered that a Pb-

rich layer is formed near the interface. Since the yield stress decreases with an increase in Pb concentration, as shown in Fig. 12, strain concentration occurs in the Pb-rich layer and fatigue life decreases.

The fatigue fractures of 36 specimens out of the 39 tested occurred near the interface, and some defects are found on the fatigue fracture surfaces. For this reason, it is felt that the dispersion of the results is caused mainly by defects.

Solder joint specimens with artificial defects were tested in order to make clear the influence of defect size on fatigue life. The relationship between defect size and fatigue strength of the solder joint specimens is shown in Fig. 14. With an increase in defect size, the fatigue life decreases rapidly, as shown in Fig. 14.

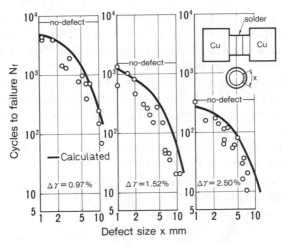

FIG. 14. Relationship between defect size and fatigue strength of solder joint specimen.

The influence of defect size on the fatigue life was investigated by a simple analysis that considers a decrease in the joined area. Assume that the shear stress on the section containing an artificial defect increases due to the decrease in the joined area. The following equation should hold:

$$\Delta\tau_c = \Delta\tau_0/(1 - \eta) \qquad (9)$$

where $\Delta\tau_c$ is the shear stress range on the section, $\Delta\tau_0$ is the nominal shear stress range, and η is the defect area rate. The relationship between the shear stress range, $\Delta\tau$, and the shear strain range, $\Delta\gamma$, is expressed by the following equation:

$$\Delta\gamma = (\Delta\tau/G) + 0 \cdot 01(\Delta\tau/\Delta\tau_y)^{1/n} \qquad (10)$$

where G is the shearing modulus, $\Delta\tau_y$ is the yield stress range at $\Delta\gamma_p = 1\%$, and n is the strain hardening factor. For the solder joint specimens tested, $G = 12$ GPa, $\Delta\tau_y = 60 \cdot 1$ MPa and $n = 0 \cdot 1793$.

Since the length of longest lives of solder joint specimens for each strain range is comparable to that of the lives of the bulk specimens, it can be considered as equal to the life length of a defect-free solder joint specimen. Equation (1) is applied to the result:

$$\Delta\gamma = 1 \cdot 06 N_f^{-0 \cdot 11} + 25 \cdot 1 N_f^{-0 \cdot 43} \ (\%) \qquad (11)$$

The defect area rate, η, was calculated from the defect size, x, and $\Delta\tau_c$ is obtained by Eqn. (9). Substitution of $\Delta\tau_c$ into Eqn. (10) gives $\Delta\gamma$, and N_f can be derived from Eqn. (11). The fatigue lives, N_f, can be obtained from the defect size, x, by the procedure described above. The calculated result is shown in Fig. 14.

Since the upper limit of the experimental values is in the range of that of the calculated values, it is considered that the fatigue life of a solder joint with defects can be estimated by considering the decrease in the joined area.

This means that the stress concentration at a defect is negligible. The reason is considered to be as follows.

(1) The strain concentration occurs within such an extremely small area that the fatigue life depends on the bulk strain.

(2) The defect edge is restrained by the hard intermetallic compounds, and the strain concentration decreases.

DISTRIBUTION OF FATIGUE LIFE DUE TO DEFECT

Since the relationship between defect size and fatigue life is known, the distribution of the fatigue life of a solder joint can be estimated from the distribution of the defect area rate.

In the fabrication of semiconductor devices, the joint surface is washed and pre-soldered, and the soldering temperature is optimized as for the solder joint specimen to decrease defects. By this procedure, the defect area rate, η, can be held under $0 \cdot 1$, but it is impossible to hold η to zero. η is distributed from 0 to 1. Though the distribution of the defect area rate in the solder layers of semiconductor devices has not been measured, in order to simplify the discussion, it is assumed that the distribution of η is given by an exponential distribution for which the average is $1/\lambda$. The probability density function, $f(\eta)$, and the distribution function, $F(\eta)$, are given by the following equations:

$$f(\eta) = \lambda e^{-\lambda \eta} \tag{12}$$

$$F(\eta) = 1 - e^{-\lambda \eta} \tag{13}$$

From Eqns. (9)–(13), the relationships between $\Delta\gamma$, N_f and F can be obtained. The failure probability, α, is given by the following equation:

$$\alpha = 1 - F \tag{14}$$

The probability distribution of the fatigue lives is obtained by this procedure. Figure 15 shows the result for $1/\lambda = 0.1$ and 0.05. A fatigue life of 5% failure probability is lengthened approximately ten times by decreasing the average defect area rate from 0.1 to 0.05, as shown in Fig. 15.

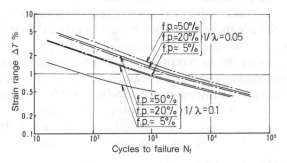

Fig. 15. Probability distribution of fatigue lives of solder joint.

OPTIMIZATION OF SOLDER JOINTS IN SEMICONDUCTOR DEVICES

The most effective means to extend the fatigue life of a semiconductor device solder layer is to decrease the shear strain in the solder layer. The strain caused by temperature difference is given by the following equation, where solder stiffness is ignored:

$$\Delta\gamma = \Delta\alpha \cdot \Delta T \cdot l/h \tag{15}$$

where $\Delta\alpha$ is the difference in the coefficients of linear expansion between two soldered parts, ΔT is the temperature difference, l is the length of the solder layer, and h is the thickness of the solder layer. $\Delta\gamma$ can be decreased by increasing h.

However, it is often difficult to control the thickness in fabricating semiconductor devices. Therefore the following methods are effective in optimizing solder joints for fatigue.

(1) In the higher strain range, the fatigue life of Pb-rich solder is longer; in the lower range, that of eutectic solder is longer. For this reason, Pb-rich solder is preferable for power devices, for example, in which a great temperature difference is caused by on/off cycles, and eutectic solder should be chosen for printed circuits where the thermal strain is not so great.

(2) The yield stress of eutectic solder is higher than that of Pb-rich solder. Thus, for a solder joint subjected to an external force caused by vibration, such as in a printed circuit installed in a car, eutectic solder is called for. In Eqn. (14) we neglected the stiffness of the solder. However, eutectic solder is advantageous because the thermal strain is lower in solder which has a higher yield stress [1].

(3) For semiconductor devices, the fatigue strength in the range below 1% of failure probability becomes a matter of great concern. In this range, the fatigue life can be extended considerably by decreasing the defect area rate, as shown in Fig. 15. Therefore the methodology of decreasing defects has become an important technical subject. Figure 2 shows an example of the extension of fatigue life by improving the solderability.

CONCLUSIONS

(1) The universal slope method was found appropriate for determining the low-cycle fatigue strength of Pb–Sn solders.

(2) In general, with an increase in temperature, the fatigue life of solder decreases. However, the influence of temperature differed with composition of the solder tested, and was found to be at a minimum for 60Pb–40Sn solder.

(3) Fatigue behavior of 95Pb–5Sn solder at 150°C was different from that of samples of other compositions and temperatures. This was found to be the result of the phase transformation of the solder.

(4) In the higher strain range, the fatigue life of Pb-rich solder was

longer; in the lower range, that of eutectic solder was longer. The relationship between composition and fatigue life in the lower strain range exhibited the same tendency as the relationship between composition and yield stress.

(5) A defect-free solder joint interface did not positively influence the fatigue strength of solder.

(6) The fatigue lives of solder joints with defects were obtained by using the defect area rate. Thus, the probability distribution of fatigue life was estimated from the distribution of the defect area rate. The probability distribution of the defect area rate had a considerable influence on fatigue strength in the lower range of failure probability.

REFERENCES

[1] S. Shida, T. Sakamoto and A. Yasukawa, *Nikkei Electronics*, No. 317, 227 (1983).
[2] K. C. Norris and A. H. Landzberg, *IBM J. Res. Dev.*, **13**, 266 (1969).
[3] P. Lin, J. Lee and S. Im, *Solid State Technol.*, 48 (1970).
[4] E. Levine and J. Ordonez, *Electron. Components Conf.*, **31**, 491, (1981).
[5] R. T. Howard, *IBM J. Res. Dev.*, **26**, 372 (1982).
[6] V. D. Coombs, *ASTM Spec. Tech. Publ.* (Am. Soc. Test. Mater.), No. 515 (1972).
[7] H. S. Rathore, R. C. Yih and A. R. Edenfeld, *J. Test. Eval.*, **1**, No. 2 (1973).
[8] S. S. Manson, *Thermal Stress and Low-cycle Fatigue*, Ch. 4, McGraw-Hill (1966).

Index